S0-BIU-001

# PLAIN TALK about the HUMAN GENOME PROJECT

## A Tuskegee University Conference on Its Promise and Perils ... and Matters of Race

EDWARD SMITH
WALTER SAPP
Editors

TUSKEGEE UNIVERSITY
Tuskegee, Alabama 36088

© Copyright 1997 by Tuskegee University

Several articles in this publication are copyrighted by the authors. This is
noted on the first page of each of those articles.

Published in the United States of America by
    Tuskegee University
    College of Agricultural, Environmental and Natural Sciences
    107 Campbell Hall
    Tuskegee, AL 36088

*Library of Congress Cataloging-in-Publication Data*

Tuskegee University Conference on the Human Genome Project: 1996 :
    Tuskegee University)
        Plain talk about the human genome project : a Tuskegee University
Conference on its promise and perils ... and matters of race /
Edward Smith, Walter Sapp, editors.
        p.   cm.
        Includes bibliographical references and index
        ISBN 1-891196-01-4 (paperback)
        1. Human Genome Project--Congresses. 2. Human Genome Project--
Social aspects--Congresses. I. Smith, Edward (Edward Jude), 1961-
. II. Sapp, Walter (Walter James), 1934-   . III. Title.
QH447.T87   1997
599.93'5--dc21                                              97-44115
                                                               CIP

# TABLE OF CONTENTS

# FOREWORD

Tuskegee University begins its 117th year this year. It has had a track record in higher education of taking on new challenges in changing times and of making a difference in the lives of people. This goes back to the founder of this institution, Booker T. Washington, and the scientist he brought to this campus, George Washington Carver, whose life's work was centered here. Through Carver's research, teaching, and outreach activities so many new uses for common things around us, mostly plants, were created as a result of his application of knowledge to the opening up of the mysteries of nature for the resolution of human problems.

The hosting of the Tuskegee University Conference on the Human Genome Project and this book which issued from it are very much in line with the vision of both Washington and Carver, albeit on a different level and on a different scale. The intensification in the use of information technology now enables us to see things and reach knowledge-based conclusions that were beyond the dreams of people of earlier times. Today we do these things on a global scale because technology makes it possible to collaborate daily.

Although we find out from the media on almost a regular basis these days of this or that gene being found, most people do not know very much, if anything, about the Human Genome Project, at least by that name. But it portends to greatly influence our individual and collective future. What are we to think about the possibility that the deciphering of the entire human genetic code is within reach? Our minds race ahead thinking of the possible alleviation of human suffering to which this could lead, and hope wells up in us. But other thoughts also crowd in. Are we going to start thinking of ourselves as nothing but genes, as determined by some biochemical configuration in our chromosomes? If we discover we have "defective" genes, what will be our assumptions about what constitutes "healthy" genes? What will alternative preconceptions do to us morally and psychologically? Is gene therapy going to be a reality and how soon? Besides these new internal threats to our psyche and physical well-being, what about the external threats of job discrimination or denial of insurance because of a genetic condition we may have? The promise and the perils of the Human Genome Project are topics well worth our attention because this machine is in high gear and, as Dr. Troy Duster says, "it will not be stopped." It is time for plain talk about the Human Genome Project.

This book also fits within the context of a new effort being planned for this historic campus and that is the launching of a Tuskegee University Center for Bioethics in Research and Health Care. What would be the mission of this Center? It will be about educating Tuskegee students relative to ethics and its application to everyday life. It will be about training programs in bioethics related to research and health care. It will be about monitoring health-related research to assure benefits for minority communities. Finally, it will also serve to preserve the memory of what happened here in Macon County, Alabama, between 1932 and 1972. I refer, of course, to the infamous "Tuskegee Syphilis Study," as it is called, in which many poor, black men from sharecropper families were used as subjects in a study of untreated syphilis by the U.S. Public Health Service. The infamy stems largely from the fact that, when penicillin was found to be an effective treatment, it was withheld for the sake of continuing the study of the course of syphilis when left untreated. The humanity, the individual preciousness of these men, was forgotten.

The publicity surrounding this study only served to corroborate among African-Americans that they were viewed as having less value in this society than other groups, and the "Tuskegee Syphilis Study" has come to stand for racism in medicine. The new Center would seek to address the challenge that science and technology presents to ethics, philosophy, theology, the social sciences, the humanities, law and medicine. Such things as cloning, in vitro fertilization, multiple births, the understandings and findings coming from the Human Genome Project, and discussions about possible genetic manipulation of human physiology or behavior or about genetic testing that could be misused—all these are current or near future happenings and concerns. Race still counts in American society. The past has some warnings for the Human Genome Project and for all of us who will live in the wake of its findings in the next century. Tuskegee University is proud to have gathered in this volume for your consideration and understanding the thoughts and experiences of a sample of outstanding active participants in the Human Genome Project—the scientists, the philosophers, the ethicists, and legal analysts watching over the common good.

Benjamin F. Payton
President
Tuskegee University

# PREFACE

In scope, potential benefit, and amount of research expenditure, the Human Genome Project (HGP) is considered by many to be one of the most important biological research projects of our times with significant implications for future biomedical and life sciences research. However, the history of biomedical research in this century, even with all of its notable contributions, has often been tainted with controversy, for example, the eugenics policies of Nazi Germany, which was contemporaneous with the beginnings of the Tuskegee Syphilis Study, to current attempts to link intelligence and criminal behavior to genetic factors. In the Spring of 1996, the idea was conceived of holding a conference at Tuskegee University whereby scientists and layman might learn more about the promise and perils of the Human Genome Project which figures to be such an important part of our immediate biomedical future. Our thinking was that an understanding of the HGP with its technological sophistication, broad social consequences and ethical concerns is essential if society is to benefit from developments generated by it. Thus the Tuskegee University Conference on the Human Genome Project was held September 26-28, 1996 at Tuskegee University with primary support of the two agencies sharing responsibility for the Human Genome Project—the U.S. Department of Energy and the National Institutes of Health. The U.S. Department of Agriculture provided additional support.

The significance of holding this conference at Tuskegee University was clear. Macon County, Alabama (Tuskegee is the county seat), was the site of the Tuskegee Syphilis Study of 1932 to 1972 in which rural, mostly poor African-American males were subjected to experimentation by U.S. Public Health Service doctors wishing to know more about the effects of untreated syphilis. These subjects were unknowingly denied an effective treatment for syphilis that was discovered during the course of the study. The study, terminated in 1972 as a result of a public outcry and a class action lawsuit on behalf of the subjects, is synonymous with scientific misconduct with racial overtones. The legacy of mistrust in medical research by the African-American community remains to this day. But, as President Clinton said on the occasion of the official apology to the study survivors (May 16, 1997): "...it is in remembering (the) past that we can build a better present and a better future....We need to do more to ensure that medical research practices are sound and ethical, and that researchers work more medical research practices are sound and ethical, and that researchers work more closely with communities." It was in that same spirit that speakers and conference participants were brought together in this Tuskegee setting to promote an understanding of the science and impacts of the HGP.

The conference was intended to meet several objectives: (1) to clarify for participants—some of whom did not have a scientific background—the goals and potential benefits of the research and introduce them to the technology being used; (2) to identify the legal, social and ethical implications of the project, particularly as they might affect the African-American and other minority communities; (3) to add an historical dimension by examining how spurious theories have been propounded in the name of science in the past; and (4) to share information as to how faculty and students at Historically Black Colleges and Universities might take advantage of opportunities to participate in genome-related research as a goal in itself and as a means of generating student interest in the field of biotechnology. The development of resource bases for genomic research at HBCUs will allow them to make their contributions to important genetic questions while providing for the training of the next generation of minority scientists in a technology considered to be one of the most essential in the medical and allied sciences.

Speakers were some of the principal researchers working on and publishing about the project and scholars and experts concerned with safeguarding the public interest as genetic information becomes more available. It was generally agreed that we had a stellar roster of invited speakers and workshop panelists. The presentations and dialogues were of such a frank and pointed nature that this proceedings could only be called *Plain Talk about the Human Genome Project.*

We call attention to readers not familiar with the science of molecular genetics or the stated promises of the Human Genome Project to two appendices. Appendix 1, *Primer on Molecular Genetics*, begins on page 265 and introduces some basics of the science. Appendix 2 follows with two case studies involving hereditary colon cancer, which illllustrate the potential of the research.

In the keynote address, Dr. Ari Patrinos, director of the Human Genome Program for the U.S. Department of Energy, gave an historical perspective on the project by tracing the origins of the project within DOE and its predecessor agencies, particularly the Atomic Energy Commission, to the tensions of the Nuclear Age and the Cold War, which required knowledge about the effects of exposure to ionizing radiation on human genes. Patrinos notes the initiation of collaborative research on the human genome between DOE and the National Institutes of Health which continues to this day. In his chapter, you will learn that the scope of the HGP goes beyond understanding the human genome. It also encompasses such needs as energy production, environmental protection and remediation, agriculture, industrial processes, clean-up of toxins. Surprisingly, the study of the genomes of a handful of experimental organisms (bacteria, yeasts, worms, fruit flies, the mouse) plays an important role in these new processes and technologies as well as in the understanding of the human genome—so similar are they in structure and function. Finally, Patrinos introduces readers to the term "ELSI," i.e., the Ethical, Legal and Social Implications of the Human Genome Project. Both DOE and NIH have programs that deal with ELSI, and an entire section of this book is devoted to these issues.

Dr. Rick Myers, director of the Stanford Genome Center at the university's School of Medicine—one of about 12 such centers, introduces Section 1 of the book: The Promise of the Human Genome Project. He clearly identifies the goal of the HGP: to find all the human genes and to know their location. He gives a couple of examples of how the finding of two genes (it is estimated that there are about 100,000 in the human genome) is enabling better diagnosis and treatment of disease. Some of the promise has already occurred, in other words, and the HGP "has already had a major impact on our understanding of ourselves, on the way we view diseases in humans, and on the way biologists practice science," according to Myers.

Introduced as one of the giants of human genome work was Dr. Maynard Olson who is involved in genomic research at the University of Washington, Seattle. His chapter provides a "window into the work that goes on within the Human Genome Centers" and how it is progressing. He rather demythologizes the technology and indeed addresses policymakers—calling for a broadening of participation by farming out 10 to 20% of the data collection to other institutions of higher education acting as satellites to existing major Genome Centers. In a postscript, he tells about just such a satellite model currently being planned between his institution and Tuskegee University.

Dr. Marilyn E. Thompson, an African-American postdoctoral fellow (NSF) at Vanderbilt University's Cancer Center, reported on the "race" to find the breast cancer gene beginning with the groundbreaking work of Dr. Mary-Claire King at the University of California-Berkeley School of Public Health which located *BRCA1* on chromosome 17. It was believed there would be other genes responsible for breast cancer and indeed *BRCA2* is being linked to chromosome 13.

Moving to more complex disorders, Richard S. Cooper, M.D. of Loyola University's Stritch School of Medicine in Chicago, describes the work of the International Collaborative Study on Hypertension in Blacks of which he is principal investigator. The study is investigating the genetic underpinnings of hypertension in three geographic regions and finding low levels of hypertension in rural West Africa, intermediate levels in the Caribbean, and high levels in the U.S. More difficult to define are the environmental and psycho-social conditions that appear to be involved in the 2:1 incidence of hypertension when comparing blacks to whites. Cooper has much to say about racism as a "fundamental obstacle to a scientific understanding of human biology."

Technology transfer is an important aspect of research programs today dealing as it does with transferring new knowledge and technologies to other commercial purposes. It used to be that scientific developments resulting from government contracts (read taxpayer money) were not available to other private efforts. Dr. Ron King heads up the office in NIH that promotes technology transfer, and he explains the turn-around.

Dr. David Botstein was the last of the speakers to address the science of the Human Genome Project, but because his presentation was of the nature of a response to the many ELSI issues raised at the conference, it begins on page 203 in the section called "Recapitulation."

As the first Human Genome Conference on the campus of a predominantly African-American campus, matters of race and diversity were sure to be addressed. Section 2 is called "Matters of Race and Diversity." It opens with a presentation by Dr. Luigi Luca Cavalli-Sforza, who is pre-eminent in his leadership of an international effort to systematically study the genetics of human populations under the aegis of the Human Genome Diversity Project. He shares some of the important conclusions of his life-long work and that of others in human population genetics. One is that races do not exist.

Dr. Raymond Zilinskas picks up on international initiatives stemming from the Human Genome Project. He notes that an increasing number of national governments, particularly those with new or well-established pharmaceutical industries, are gearing up to participate in human genome work. He goes into some detail in describing genomic research efforts and structures in China, Japan and India. He predicts that the new knowledge generated by the project will translate into many new products and therapies as we move into the 21st century.

The promises of the HGP are profound, says Dr. Patricia King, insofar as the medicine and biology derived from it may alleviate human suffering. But the perils too are great. She fears that the complexities of societal problems will be treated simplistically via genetic explanations and that minorities—and the poor in general—will, as some have warned, find themselves relegated to a biological underclass.

In his presentation on "The Responsibility of Scientists in the Genetics and Race Controversies," Dr. Jon Beckwith points out that science is not a neutral activity in itself and that scientists like the rest of the population bring their personal biases to their work. He offers a number of examples of the misrepresentation of genetics in the media and the current "revival of racialist science." Further, he argues for the participation of scientists in the public discussion of new technologies that involve significant societal impact.

As an African-American anthropologist, Dr. Fatimah Jackson criticizes the HGP's limitation of its baseline reference panel to a small number of "opportunistically available cell lines" representing only North Atlantic European-Americans. She calls for representation by other older cell lines.

Dr. Georgia Dunston follows up on the criticism and outlines Howard University's program of Genomic Research in African-American Pedigrees (G-RAP) envisioned by her and her colleagues. In 1995, the Human Immunogenetics Core Laboratory at Howard began three collaborative projects—on mutational analyses of BRCA1 in high-risk African-American families, West African origins of NIDDM (non-insulin dependent diabetes mellitus) in African-Americans, and the genetics of prostate cancer in African-American men. These projects are foundational to their G-RAP plans.

Attorney Hans Goerl introduces Section 3 called "The Perils of the Human Genome Project" which addresses the HGP's ethical, legal and social implications and genome data use. He clearly points to the dangers—the money (power) to be made (controlled), the impact on society and the individual, and the historical precedents set by the eugenics movement in getting rid of "socially undesirables." He also spoke about genetic discrimination in employment, a topic not covered by other presenters.

The presentation by Dr. Courtney Campbell of Oregon State University puts a human face on the many concerns surrounding the Human Genome Project, and for this reason it was placed early in the Perils section. He tells the story of his friend who may be at greater risk for breast cancer. Should she try to find out?

Elizabeth Thomson was the head of the Ethical, Legal, Social Implications Research Program at the National Human Genome Research Institute at NIH. She reviews the origins and goals of the ELSI programs, particularly at NHGRI, and identifies the four major issues that have emerged. This presentation is important because the ELSI program is a funding source for various activities and allows accessibility to policymakers. Thomson added developments since the conference including a policy that seeks to protect the genetic privacy of DNA donors by ensuring informed consent and that "the initial version of the complete human DNA sequence is derived from multiple donors from diverse populations…" Several at the conference were critical that HGP work had been based on DNA taken only from donors of European-American descent. Importantly, Thomson reports, "clarifications are still occurring."

Another presentation that needed to come toward the beginning of the Perils section was that of Dr. Philip Bereano. Clearly, his was one of the most provocative of the entire conference. He questions the very underpinnings of the project, calling the HGP "a prime example of how power works." He says new technologies are never neutral. They represent some group's interests and goals. The questions are: "whose goals, what goals, and … is this really the most important biology question of our times?"

Dr. Robert Murray, Jr. explains the difference between genetic testing and genetic screening and raises some of the issues involved in both. He speaks to the main ethical values and offers guidelines for both types of programs. An important part of his chapter is his account of what took place in the 1970s in what he calls "The Sickle Cell Debacle."

Genetic enhancement—using genetic testing and modification to "improve" individuals—is, according to Dr. Glenn McGee of the University of Pennsylvania, going to be inevitable. Indeed, he says, the media constantly plays up genetic enhancement and gene therapy to be the real promise of the HGP while almost totally ignoring the scope of the problems involved. He says that the lessons of parenthood are sufficient to provide guidance through an enhancement future—and we would be wise to follow them.

Through a set of scenarios, Dr. Thomas Murray takes the reader through the question: how can a society based on equality (so to speak) be just in the face of human differences that the HGP is making ever more clear (in matters of health and disease). By comparing ways of distributing other goods (merit and purchase), he addresses the issue of a just system of access by all to health care. He (and an ELSI Task Force) rejected the actuarial fairness concept of the insurance industry in favor of the sole moral imperative of *need*. But David Christianson and Joan Herman, both presenting as actuaries, defend the concept. They have their scenarios too. The reader can weigh the arguments.

It is generally recognized that present laws will not protect the information that will be generated as a result of the Human Genome Project. Patricia (Winnie) Roche shared the proposal of a group from the Boston University School of Public Health for a national Genetic Privacy Act. By 1996, this proposal and the "Genetic Confidentiality and Nondiscrimination Act of 1996"(S. 1898) which Roche says is based on the GPA, were the only serious efforts at writing legislation protecting individual privacy. But the Clinton administration backs S. 89 (and H.R. 306), the Genetic Information Nondiscrimination in Health Insurance Act of 1977, which Roche considers too narrow as it focuses primarily on uses of genetic information in health insurance and ignores other issues.

Dr. Vijaya Melnick rounds out the Perils section with a plea to reconsider our concepts of normality, abnormality, disease and disorders. She introduces two more human faces—both professors, one autistic and the other struggling with manic-depressive disease. Both have shown great courage and creativity in their lives and would not choose to be without their struggles. Food for thought is this presentation that fears we may be moving into an Orwellian future.

The final chapter in Section 3 is titled "Who's That Fish on My Line?" Again, we hear from Hans Goerl. As an editor for a listserve for geneticists, he knows from whence he speaks. He reveals the thrust of his presentation with his subtitle: "The Dangers of Electronic Distribution of Genetic Information." Because of the availability of huge volumes of new information—much of it being interpreted on the Internet with no guarantee of trustworthiness, this is a must read for net users and a plea for restraint and sensitivity on the part of information providers.

The section called "Recapitulation" needs to be read. Dr. David Botstein believed that an ELSI tide had washed over the conference. While the program book had a subtitle of "Social, Ethical and Legal Implications," the meeting strove to give a fair hearing to both sides of the Human Genome Project—or, as we have been terming it, the promise and the perils. Botstein says he is "one of these guys who believes in the weighing of the good against the bad," and so he attempts to clarify the limitations and the potential for good in the HGP. He is particularly proud of the "infrastructure...of roads on which an industry of genomic trucks can ride efficiently and provide all kinds of useful information." He is talking about the large-scale identification of disease genes, and in his conclusion he makes a very strong case for the HGP.

Dr. Troy Duster's presentation likewise was a reflection on the thoughts expressed in the conference sessions and indeed to the conference itself. He identified "fuzzy thinking" on the part of both the so-called "ELSI types" and the molecular scientists and provided some very useful insights, especially on the science of race and what he (following Felix Kaufman) termed first order and second order constructs. While the HGP has had little connection to the African-American community, according to Duster, African-Americans have a big stake in the HGP not only because of this group's past experience in this country but also given the recent propensity for genetic claims of complex behaviors. Herein lies, he says, the relevance of the conference at Tuskegee University.

Section 5 speaks to educational aspects of the Human Genome Project. The section begins with a concept paper offered by Drs. Robert Yuan and Spencer Benson of the University of Maryland, College Park. Noting that universities have been slow to meet the challenge of preparing their undergraduate and graduate students in biological sciences for an increasingly complex and changing world, Yuan and Benson offer a useful plan for providing diverse student bodies with the processes and skills needed for scientific understanding or careers.

Dr. Paula Gregory, chief of Genetics Education at the National Human Genome Research Institute at NIH, describes opportunities available through her office: The Genetic Self Workshop, DNA sequencing for high school students, and a short course for faculty at minority institutions. Interested parties can use this chapter as a resource.

The next four presentations were given at a panel discussion on *Promoting Awareness of the Human Genome Project*. Dr. Virginia Lapham reports on the Human Genome Education Model, a joint project of Georgetown University Child Development Center and the Alliance of Genetic Support Groups. They want to develop an education model for those with a genetic disorder (consumers) and for the genetic services providers. She reports the results of a national survey of 332 consumer and 329 health professional respondents as to their priorities in educational topics.

Aleta Sullivan, now an instructor at a community college, tells of her 13 years of experience as a high school teacher of biology in Mississippi. She has availed herself of a number of opportunities for educating herself in molecular biology and has become comfortable with making the most of the Internet in her classroom instruction. High school and college faculty will profit from reading this chapter.

Carlton P. Jones, a 1995 graduate (M.S. in Animal Science) of Tuskegee University, speaks as a student. He currently is pursuing a Ph.D. at Michigan State University. He shares his thoughts on the need to expose high school students (college students too) to the various fields of science that are open to them. He also calls for many people becoming aware of the HGP because of its implications for the future.

Dr. Edward Wheeler, a recent Dean of the Chapel at Tuskegee University, speaks for the religious community in general and the African-American Church in particular as he addresses the role of the Church in increasing community awareness of the HGP. He notes the suspicions that have existed historically between the scientific and religious communities and presumes a conservative stance by many churchmen and women who will be protective of their church's interpretation of the Bible's creation account. However, he believes mutual respect will open doors of cooperation particularly since the wonder and beauty of God's creation seem to be confirmed by the discoveries of the Human Genome Project.

There are three Appendices attached. Appendix 1 is taken from the U.S. Department of Energy's website. It is called a *Primer on Molecular Genetics*. As was said above, some readers, it is assumed, will not be familiar with the science of the Human Genome Project. It is hoped that the text of this appendix and its figures will provide somewhat of an introduction.

Appendix 2 comes from the National Institutes of Health website. Two briefly-stated case studies on hereditary colon cancer show the power that genetic tests are going to exercise over our lives more and more as DNA sequencing progresses. The text also points to the many issues that will need resolving. Some readers will want to start by reading Appendices 1 and 2.

Appendix 3, Useful Websites, is meant to serve as a resource. They are a source of links to other useful sites.

In conclusion, strategies to educate the public on the broad themes of the Human Genome Project and its program of work are essential in reducing and allaying public concerns and misconceptions and also in highlighting the very real concerns about possible new forms of discrimination and ghettoization of minorities. It is hoped that this volume will serve as an important addition to the literature, and that it will find its way into secondary and collegiate classrooms and the libraries of educators, researchers, students, health care providers, community leaders and other potential beneficiaries and users of genetic information produced by the HGP.

For a final reflection on this collection of readings on the promise and perils of the Human Genome Project, we call attention to Thomas Murray's statement:"...those things which are humanly most important to us—courage, creativity, loyalty, and the like—are the things we are least likely to have genetic explanations for." Those things spring forth from our human personhood that makes each of us one of a kind, worthy of equal respect, and—quite beyond any attributes that come from our genes or the environment in which we developed—never fixed but always capable of being in new ways and doing new things.

Edward Smith
Walter Sapp

## ACKNOWLEDGMENTS

We are extremely grateful for the financial help received from the U.S. Departments of Energy and Agriculture and the National Institutes of Health which enabled us to keep registration costs to a minimum and to support the participation of students from local high schools and various universities. Thanks also to Research Genetics, Inc., PE Applied Biosystems, and the Tuskegee Area Health Education Consortium, Inc. for their support. We are especially grateful to Dan Drell of DOE and Bettie Graham of NIH for assistance in putting together the program. To the distinguished speakers, thank you for providing such a rich and provocative forum that allowed for a very useful exchange of ideas.

Clearly, the success of the conference was due to the efforts of the Animal Genetics students at Tuskegee University, especially Chavonne Hubbard, Sophie Ramlal, Shakura Haqque, Youqiang Song and Sam Nahashon and their tireless work in sending mailings, transporting participants, transcribing tapes, etc. Special thanks to faculty and staff at the College of Agricultural, Environmental and Natural Sciences (CAENS) and the George Washington Carver Agricultural Experiment Station, and to colleagues at the College of Liberal Arts and Education, the College of Veterinary Medicine, Nursing and Allied Health and other units on campus all of whom worked to make this historic conference a success. Dean Walter Hill, Conrad Bonsi, Louise Herron, Gliss Pimental-Smith, Robert Zabawa, Glenn Malone, Jiliang Chiu, Connie Price, J.H.M. Henderson, Ralph Noble, Aurea Almazan, Roger Hagerty, and J. J. Johnson III are some of them. Also, we need to mention the support of the Center for Biomedical Research/RCMI and JoAnn Crooks, the Center's administrative assistant.

Finally, while the conference was our product, this proceedings represents the creative and attentive effort of Marie Loretan, our in-house editor; for this we are grateful. She in turn thanks Barbara Dunn Harrington, Lynn Ballard-Siaway, Helen Benford, Edith Powell, Phil Loretan, Mary Gaines, Tom and Shirley Curran, Marshall Peterson, Betty Mansfield and co-workers, Glenn McGee, Garth Green, and all the authors and their staffs for their cooperation and help.

It is our hope that this book will contribute to the debate about the Human Genome Project and to define the range of opinions that exists about this very important biomedical research program.

The Editors

# THE HUMAN GENOME PROJECT: WHAT IS IT?

## Ari Patrinos

The excitement associated with the Human Genome Project is overwhelming. We at the Department of Energy are very humbled by the great challenges this program presents to us. What I would like to do is describe the program in some very general terms. First of all, I need to answer the question of how come the Department of Energy is involved in something like the Human Genome Project in the first place. It is a very good question. I also would like to convey to you the promise that this program offers in terms of future benefits, including those important to the Department of Energy. I will also cover some of the accomplishments of the Human Genome Project, and I will be unabashedly parochial by stressing those accomplishments to which the Department of Energy has contributed. But I want to make it very clear that we are but one player in a fairly international program. I will end with a few aspects of the ethical, legal, and social implications of the project.

As a point of reference, I would like to say that a unifying fact of human genetics is that all of us—man or woman, African-American, Asian, Native-American, Caucasian or whatever—have genomes that are fundamentally 99.9% identical in sequence. The genome is the complete set of the genetic instructions that are locked up in the 23 pairs of our chromosomes. The DNA molecules that carry these genetic instructions are essentially linear molecules; they are made up of four simple bases; there is a shorthand for them—A for adenine, T for thymine, C for cytosine, and G for guanine. A pairs with T and C pairs with G in the well-known DNA double helix, and this pairing is at the core of human genome work and contemporary molecular science.

Although each one of our cells contains about 6 billion base pairs of DNA, 3 billion from each parent, our differences are determined by only about 1 in 1000 of these base pairs. At this fundamental level of molecular definition, it should be obvious to all of us that what unites us dwarfs anything that separates us or distinguishes us.

The author is the Director of the Human Genome Program for the U.S. Department of Energy as well as its Associate Director for Biological and Environmental Research, Germantown, MD 20874-1290.

The objective of the Human Genome Project is to map the complete DNA sequence in a typical human cell and to supply the information that will, among many things, lead us to understand the critical differences that make us individual and that could sometimes, when malfunctioning, lead to disease. That information will essentially create opportunities both to prevent and also to treat diseases.

Long before there was a Human Genome Program, the Department of Energy and its predecessor agencies, like the Atomic Energy Commission, were very interested in developing methods to detect genetic changes from exposure to ionizing radiation. Back then, it was very clear and obvious that the nucleus was in fact the information-containing part of a cell and was the most sensitive to ionizing radiation and other pollutants. As we became more capable in working with and understanding genetic material in the late '80s, we started thinking about sequencing the whole human genome. As we contemplated the enormity of the task, we recognized that the unique capabilities and resources of our national laboratories would be critically important to this particular undertaking. Our laboratories, after all, bring to bear specialized resources such as high-skilled engineering, manufacturing, and high speed and high performance computing. They also provide unique atmospheres for collaboration among the many scientists that are important in the undertaking of the Human Genome Project and with programs related to it in chemistry, physics, engineering and of course biology. The labs foster an environment where molecular biologists, chemists, biochemists, physicists, engineers, and computer scientists can work together because this project requires this mix of disciplines and sciences.

Other unique assets that the Department of Energy brings to bear are the synchroton light sources and the neutron sources that have spawned the relatively new science of "structural biology." Structural biology, using these sources combined with computational biology—another emerging revolution in the field, essentially addresses understanding the three-dimensional structures of various biological molecules, such as proteins, for example, and complements very constructively the Human Genome Project.

In terms of history, the Department of Energy committed its first funds for human genome research in 1986, and the National Institutes of Health started their own genome program in 1987. What usually happens in bureaucratic environments is that departments get together and sign something called a "memorandum of understanding"—bureaucratese for agreeing to work together, not fight each other, and coordinate individual resources and capabilities so that essentially the total effort is the sum of the individual parts. I am pleased to say that in this particular case this collaborative effort has been very successful.

The official clock for the program was started October 1, 1990. At that time the NIH and DOE issued a five-year plan to describe the goals and objectives of the program, and this is one government program that actually moved faster than what had been originally expected. It may be the first; I don't know of any other government program that can claim that. Whether it will ultimately be that way to the very end remains to be seen. But the principals in the program had to get together and revise the five-year plan because essentially some of the successes had exceeded expectations. It has only been a couple of years since another five-year plan was issued.

A term that has been used sometimes with the Human Genome Project is the "Holy Grail." I do not particularly like that term, but what is being sought in this instance is completing the 3 billion base pair sequence in the human genome—essentially identifying and cataloging it. This knowledge of the complete sequence will enable us to identify and better understand the approximately 100,000 genes that are in the human chromosomes. This information will enable studies about how each one of these genes functions and how disease happens when one or some of them malfunction.

In some sense the sequencing that I am referring to is the easy part of the job. It is like setting up a canvas for an exquisite painting. While the Human Genome Project will describe the human genome in molecular detail, its longer term impact will be to reveal critical mechanisms of human biology and supply the context for the medicine of the future. This will be a medicine that is rooted in mechanistic understanding and will involve interventions that are more targeted and more effective. This will no doubt lead to cost-effective medicine since prevention is always much more effective and certainly much cheaper than treatment.

The promise of the HGP includes many things. I would mention health-risk assessment, precise disease diagnosis and, as I have already mentioned, better therapies. The promise of the program, however, extends beyond just medicine to other areas of science. Allow me to spend a few minutes describing some of those.

The program has fostered the study of genomes of other organisms and this will have many useful applications in areas such as energy resources, environmental protection and remediation, agriculture, better materials and cleaner industrial processes. One can envision new varieties of plants, for example, for renewable biomass-based energy production; this is another type of research which DOE supports. Biological catalysts, such as enzymes and catalytic antibodies, could be designed for mining and processing the way one designs, for example, the mechanical components of industrial systems.

Bioprocessing will minimize pollution while bioremediation will clean up wastes. As you know, the Department of Energy is in charge of cleaning up many wastes, particularly those associated with the 50 years of the Cold War. There are many sites around the country that have been contaminated by the by-products of weapons production. It is reasonable to expect that the biotechnology of the future that will be based on genomics will in fact accomplish—and here the term "Holy Grail" is appropriate—society's objective of the sustainable development of the future. In our program and related to this, we are very proud to have started what we refer to as the Microbial Genome Initiative. In this effort we are characterizing the genomes of microbes that are of interest from energy, environmental and industrial perspectives. Although many micro-organisms, as you know, are pathogenic, many others are extremely valuable agents for many important functions. Some microbes live and thrive at boiling water temperatures or at pressures that are found only in the deep ocean. Others are found in extremes of acidity and alkalinity and many just simply love to consume things that are very toxic to humans and other living things. By developing an understanding of the genes and the associated proteins that enable these microbes to survive and prosper under these harsh environments, we will harvest the inherent potential of such organisms and perhaps train them to help in catalysis of industrial processes or to accelerate a clean-up of contaminated sites. I would like to mention some of the success stories.

Gary Saylor, a scientist at the University of Tennessee, is applying *Pseudomonas aeruginosa* in the bioremediaton of sites contaminated by naphthalene such as at our Oak Ridge National Laboratory site, a place where I spent five years of my career. Another example comes from the completion of the sequence of *Mycoplasma genitalium* and *Methenacoccus jannaschii*, work that was done by Craig Venter and his colleagues at the Institute for Genomic Research in Gaithersburg, Maryland, just down the road from our headquarters. Last year, the consortium of American, European and Japanese laboratories completed the 12 million base pair sequence of yeast—brewer's yeast. As you know this organism is a eukaryote with many structural and functional similarities to the typical human cell. There are many other examples that have been funded by other agencies.

The study of these simple cells and simple organisms helps us understand human cells. In fact, the genes that determine structure and function for simple cells are often quite similar to those that determine structure and function for human cells. I said earlier that not much distinguishes us as human beings. I can also say that not much distinguishes us from other living things, at least at the fundamental, molecular and cellular level. That has been one of the most exciting and unexpected discoveries in this journey of exploration—the fact that all living things share so many things.

I would now like to highlight—as I said, from a parochial point of view—some of the contributions that the Department of Energy program has made to the Human Genome Project. Our national laboratories, particularly Los Alamos National Laboratory and Lawrence Livermore National Laboratory, have been very instrumental in some of the highest resolution physical maps, particularly of chromosomes 16 and 19. They are now the benchmarks for some of the work that lies ahead in terms of large-scale sequencing. A major principal investigator that we have at the University of Washington, Leroy Hood, has sequenced nearly a million base pairs of DNA from both the mouse and human T-cell receptor complexes. This work has provided insights into the function of certain white blood cells, important for many immune responses to invading pathogens and thought to be involved in both auto-immune diseases and protection against early tumor development.

Another one of our contributions has been the development of novel vectors that are critical for the manipulation of DNA in fragment sizes that can be readily characterized and studied. One of those is referred to as the "BACs" which stands for bacterial artificial chromosomes. One of the BAC libraries in fact contributed to the two recently discovered breast cancer genes that are much publicized in the media.

Another major challenge for the Human Genome Project is how to deal with the avalanche of data that will result from large-scale sequencing. Both the NIH and the Department of Energy firmly believe in making all that data publicly available in the shortest possible time after the data are acquired and quality assured. Furthermore, it is imperative that access to all the data be extremely user-friendly so that all scientists can use them in their research pursuits. Many analogies have been used to highlight the size of the database that will include the complete sequence of the human genome. One is that, if you just print the three billion A,T, C and Gs, they would occupy over 200 large city phone books. This is not an enormous database as compared to databases that we have used, for example, in some of the global change studies that involved satellite observations from space. But nevertheless the human genome database will still be of great enough size and complexity that it will require serious thought in terms of how to deal with it.

From the beginning the human genome program has been a highly-focused but constantly evolving enterprise. Right now the program stands at the threshold of embarking on the serious job of getting down to large-scale sequencing. Consequently, we expect that a large part of the resources of the program will be devoted to this production activity.

Our principal goal remains the complete sequencing of a generic human genome by the year 2005. The job will by done by many laboratories and universities in this country and abroad. Even before this major job gets underway, we are already exploring how we need to prepare ourselves as scientists and science managers to deal with the challenges of what we have been calling the "post-genome world" or "Genome, The Sequel" and many other perhaps not very imaginative terms. I like to call the period "After the Revolution" because this program, I firmly believe, will revolutionize biology.

I also need to say that this wonderful community of biologists involved in the Human Genome Project will have to undergo major transformation to take on the challenge of larger scale sequencing which requires more of an engineering and production mentality. That is perhaps alien to a community which is composed mainly of small scientists, and I do not mean small in a derogatory sense. That is the observation of somebody who has infiltrated the field from outside and perhaps has had brushes with what is referred to as big science, be it laser fusion, high energy physics, nuclear physics, and—in the last decade or so—global environmental change research.

I now come to the last part of my talk, the part that I am convinced is the most important. In fact one of the most significant contributions of James Watson who, as you know, won the Nobel Prize for discovering the structure of DNA, the double helix, was his recognition that knowledge derived from genome studies had broader medical and social implications.

This led to a sub-program that was supported by the NIH and DOE, that is referred to as ELSI, the Ethical, Legal, and Social Implications of the Human Genome Project. The goal of ELSI is to address the implications of vastly increased genetic information to individuals and society. Both in the U.S. and abroad, genetic information has been misused in the past. So we need to make absolutely sure that such mistakes are never repeated.

The ethical, legal, and social issues associated with the genome program are many. I will just give you a few from the long list.

1. The fair use of genetic information;
2. The impact of genetic testing;
3. Impacts on personal reproductive decisions;
4. Misuses of genetic information;
5. Privacy implications of personal genetic information in various settings, e.g., the workplace, schools, adoption practices, and other such areas;
6. Issues of commercialization and intellectual property of genome results; and
7. Implications of personal genetic variation.

Let me also mention genetic literacy particularly in dealing with complex conditions that involve multiple genes and gene-environment interactions where the science is still poorly understood. Clearly, informing a woman that she carries a gene associated with a high risk of developing breast cancer is a serious issue. This is especially the case if options for treatment are difficult, painful, debilitating and often unsuccessful.

Recent studies suggest that many women in high-risk families simply prefer not to know. On the other hand, there are others that very much want to know such as insurers and employers. As you know, there are a number of bills before Congress that are attempting to address these issues, particularly dealing with the confidentiality of genetic information.

Let me mention some of the successes of our ELSI program. Developed by one of our ELSI grantees, a model genetic privacy bill was developed, and parts of it have been incorporated into the Genetic Confidentiality and Nondiscrimination Act introduced by Senator Pete Domenici in June of this year. Through ELSI, we have also sponsored workshops on genetics and the law. I was able to participate in one of those last July; it was a wonderful experience. Those workshops help judges to better understand and appreciate the relevance of genetic information in court.

Dan Drell of our DOE staff has been very capable in spearheading the ELSI efforts for the Department of Energy. He has been very instrumental in producing curricula for high schools that impact approximately 2.5 million students of high school biology. Many of those students will take that course as their last exposure to scientific endeavors.

Another ELSI project is exploring the implications of patenting genome sequences and the transfer of genome information and technologies to the commercial sector, a subject of considerable controversy and one that will be hotly debated over the next few years as the products of genomic research become more and more prominent in the marketplace.

I have to say that many of the ELSI issues we are dealing with are not new in medicine but will become much more acute with the progress of the Human Genome Project. Our challenge is to anticipate these problems, analyze them, and encourage healthy debate about their resolution and better inform policymakers and the public. This is one case where we cannot err on the side of too much information.

We see many ELSI challenges confronting us in the future. Informed consent, for example, for participants in genetic research will remain an important issue. Genetics involves shared familial information and the diagnosis for one person has direct implications for his or her family members. It is extremely important that patients and research participants fully understand what information and future predicted insights about them emerge from genetic studies, particularly when they involve genetic testing and screening for multi-genic and predisposition factors. What do you tell someone who tests positive for a disease-associated defect when you can only be vague about its clinical implications? What responsibilities do physicians and counselors have in the communication of risk information to a patient when the risk itself is in fact poorly understood?

Looking ahead to the new century, we see a major paradigm shift for biology and medicine. I often tell the high energy physicists and nuclear physicists, my colleagues in the Office of Energy Research, that their time has come and gone, and that the next century belongs to biology.

The completion of the human genome sequence will promote myriad new insights into the fundamental processes that govern our lives and open new possibilities for disease prevention and cure. Along with the benefits, we will have to deal with all the associated ethical, legal and social issues. How successful we are in dealing with these issues will in fact determine how worthy we are of the expected dividends of genomic research. I hope that the presentations offered at this conference will contribute, if even just a little bit, to the success of this endeavor.

# 1

# The Promise of the Human Genome Project

# THE PROMISE OF THE HUMAN GENOME PROJECT: AN INTRODUCTION

## Richard Myers

Before going into the science of the Human Genome Project, let us first spend a little time speaking about the essentials of the HGP. The way that most scientists view the goals of the project is that we want to find all the human genes. That encompasses a lot of different technologies, and a lot of different issues are raised by the very loaded concept of finding all the genes in the human organism. Not only do we want to find them, we also want to know their location along the length of the chromosomes.

In addition, we are studying the genomes of other organisms—the *E. Coli* bacterium, yeast, fruit fly, a nematode worm, and the laboratory mouse. Some of these genomes are already complete. There are many reasons why this work is important. In particular, complete genome sequences of these "model" organisms interpret and understand biological processes and their control. In addition, the identification of genes in the human genome is aided tremendously by having the sequences and other biological information from other organisms.

Another major goal of the Human Genome Project is to figure out how to store and disseminate this information. Many of the architects of the HGP and others who have become involved in it have been worried about this problem, and we still do not know very much about how to deal with it. We are still learning a lot. We have only scratched the surface, compared to the amount of information that will be generated over the next ten years.

Another goal of the project—not listed explicitly in the five-year plan—is that we would like to understand the variation in the human genome, how the genome differs in DNA sequence among different populations. There are many ways of interpreting this statement. One of them is that it will allow us to find genes and the causes of genetic diseases. It will also help identify genes that play a role in other traits. That is where much of the controversy and the concerns on how this information will be used arises.

The author is Professor of Genetics and Director of the Stanford Human Genome Center, Stanford University School of Medicine, Stanford, CA 94305-5120.

From the beginning, genome scientists and other biologists, aware that this information was coming, have stated that the project is about gathering information. The next question is: how do we actually interpret it? The genome project's goals do not state how this will be done. We are talking about identifying 100,000 genes. Those who work with genes know that every gene becomes an industry in the research world today. Many laboratories spend much of their time sorting out how even single genes function and what role they play in biological processes.

We suddenly have huge amounts of information. We are going into a phase when researchers will not need to hunt for genes but will need to figure out how to put them together and understand their functions. Most biologists realize that it would be a gold mine if years of searching for the key reagent to understand a process can be bypassed. The Genome Project is changing the way research is done in biology and will have a much more dramatic effect when all human genes are identified.

Another major part of this interpretation is the interplay between genes and the environment—in the case of humans, we would say genes, the environment and cultural contributions—and how that interplay contributes to an overall phenotype. This is really where the difficulty lies. It would be arrogant for any geneticist or any other biologist to say that we are going to be able to figure all that out. It is going to be a hard and very challenging problem. Some of the presentations in this symposium will remind us of that interplay. It is not all genes; it is not all environment. It is clearly both.

Another issue is that human geneticists think of genetics in a lot of different ways. To give an example, there are simple genetic defects that follow so-called Mendelian inheritance. They look like they are due to mutations in single genes. Cystic fibrosis and sickle-cell anemia are examples of this type of genetic disease. However, they are hardly simple when it comes to understanding them in depth. Some people have a mutant genotype for such a disease, but they have a mild or even non-existent phenotype. So trying to figure out how the mutations in these simple diseases cause the phenotype is often not a simple process. This becomes much more complicated for other genetic disorders, particularly ones that have contributions from multiple genes.

There are more than 3,000 single gene traits described for humans. Many of these are very rare, although in aggregate they have a significant effect on human health. Some of these are ones that are familiar—Huntington disease, cystic fibrosis, colon cancer, etc.

Other diseases, the ones often referred to as "complex genetic diseases," also have a genetic component. In fact, they could be entirely genetic in some cases requiring mutations or variation in multiple genes. They are multi-genetic or complex diseases. There are many layers of added complexity. As Dr. Cooper says in his presentation, you may have a common variation in a gene that is present in lots of people, but it requires common variance in multiple genes and perhaps interactions with the environment to get the phenotype. One reason why there is more focus being placed on these complex disorders is that, while there are thousands of single gene disorders out there, many of those genes are being found. The Human Genome Project will clearly find most of these. It is the other ones that are really difficult and that are much more common—diseases like heart disease, autoimmune diseases, certain psychiatric diseases, many cancers. They have multiple gene as well as environmental components.

Let me give two examples of what is going on in the Human Genome Project to find human genes. The first one is the case of the gene responsible for the most common form of skin cancer, called basal cell carcinoma. Dr. Matthew Scott, a *Drosophila* biologist at Stanford University, was studying a gene called "patched" for several years. It plays a role in early development in fruit flies. He, his graduate students and his post-docs found the human version of that gene. You have heard about conservation of DNA sequences across species that is a consequence of the fact that all living organisms are derived from common ancestral organisms in ancient history. Even with this conservation of functions and DNA sequences, it is often hard to make the jump and find versions of the same gene in species too far apart. Nevertheless, members of Matt's laboratory were able to isolate the mouse and then the human version of the patched gene. David Cox and I mapped the gene at our genome center and found that it is located on chromosome 9. We looked up diseases that were mapped near this chromosome, and one was basal cell nevus syndrome, an inherited disease that has skin cancer and other symptoms. Matt and his colleagues thought this might be a possible candidate. They looked through the sequence of the human patched gene and found mutations in people who have spontaneous basal cell carcinomas, as well as in people who have basal cell nevus syndrome, the inherited disease.

One reason this is interesting is that there was 10 years of biology research on the patched gene and the biological pathways in which it participates. We already knew the function of this gene because its function had already been worked out in fruit flies. This is why the Human Genome Project studies model organisms. There is a huge amount of information about many genes in these organisms and, because of the principle of conservation across species, we are able to transfer that knowledge and generate new information of medical importance to the human species.

The second example is about a gene that is involved in a type of epilepsy. Epilepsy is an extremely common disease in humans; about 3% of the world's population is affected by it. It is very heterogeneous and has many different causes, both genetic and environmental. But there is one form of it that we were studying called EPM1, or progressive myoclonus epilepsy, which is a rare inherited epilepsy with additional symptoms. Figure 1 gives insight into the value of the Human Genome Project. We found this gene by using the genome project and a process called "positional cloning." We found mutations in EPM1 patients that encode cystatin B, which is a protease inhibitor, providing evidence that this gene is the major player in this disease. One reason why this is impor-

Fig. 1. The nucleotide sequence of the human cystatin B gene, which has been shown to be involved in progressive myoclonus epilepsy. Upper case letters denote the exon sequences, and lower case letters correspond to introns. The upper case single letters correspond to amino acids. Note two different positions where single base mutations were identified in patients with the disease (labeled "splice site mutation" and "nonsense mutation"). Both changes are predicted to have detrimental effects on the expression of the protein, and indeed, it was found that messenger RNA levels are decreased from both alleles (Pennachio et al., 1996). These data, combined with the fact that the mutant alleles were not seen in several hundred randomly-selected individuals, provided strong evidence that loss of function mutations in the cystatin B gene can lead to this form of progressive myoclonus epilepsy. After the initial study was published, additional mutations were identified in other families with the disease, providing further evidence of the association of the disease with the gene.

## Cystatin B Mutations in Progressive Myoclonus Epilepsy

```
ACGTGACCCAGCGCCTACTTGGGCTGAGGAGCCGCGCGGTCCCCTCGCCGAGTCCCTCGCCAGATTCCCTCCGTC

CCGCCAAG ATG ATG TGC GGG GCG CCC TCC GCC CAG CCG GCC ACG GCC GCC GAG ACC CAGC
         M   M   C   G   A   P   S   A   T   Q   P   A   T   A   E   T   Q

AC ATC GCC GAC CAG gtgggtgggccgcggggacgggccggccggagtcctgccttagc..1364 bp ..
H  I   A   D   Q

                                    splice site mutation
gagacatcctcattctgtccctctgtcta [c/g] GTG AGG TCC CAG CTT GAA GAG AAA GAA AAC AAG
                                     V   R   S   Q   L   E   E   K   E   N   K

AAG TTC CCT GTG TTT AAG GCC GTG TCA TTC AAG AGC CAG GTG GTC GCG GGG ACA AAC TAC
K   F   P   V   F   K   A   V   S   F   K   S   Q   V   V   A   G   T   N   Y

TTC ATC AAG gtagagtgtgggcctcaggagggctg...266 bp...agaggcttcgctcactccgctctcttc
F   I   K

                                    nonsense mutation
ccag GTG CAC GTC GGC GAC GAG GAC TTC GTA CAC CTG [T/C] GA GTG TTC CAA TCT CTC CCT
     V   H   V   G   D   E   D   F   V   H   L           R   V   F   Q   S   L   P

CAT GAA AAC AAG CCC TTG ACC TTA TCT AAC TAC CAG ACC AAC AAA GCC AAG CAT GAT GAG
H   E   N   K   P   L   T   L   S   N   Y   Q   T   N   K   A   R   H   D   E

CTG ACC TAT TTC TGATCCTGACTTTGGACAAGGCCCTTCAGCCAGAAGACTGACAAAGTCATCCTCCGTCTACC
L   T   Y   F   *
```

tant is that people who had this rare form of epilepsy fared very poorly when they were treated with Dylantin, a drug commonly used successfully to treat most epilepsies. So this is another example of the ways in which the information that comes from the Human Genome Project can provide significant help to patients suffering various diseases. We will be able to stratify them and make differential diagnoses so we can use drugs that will help rather than rely entirely on the standard hit-or-miss approach.

This particular form of epilepsy is not trivial to diagnose on the basis of symptoms alone. The disease is rare and may not be the best example, but one of the hopes of the Human Genome Project is to develop the therapeutics on the basis of this information in cases where particular types of genes give you particular types of phenotypes. I am not saying that we can do this now, but it is certainly one of the motivations for the Human Genome Project. We have to be careful about the interpretation of genotype and how that gives you the phenotype because it will not always be straightforward. But that is one of the goals.

## References

Pennacchio, L.A., A-E. Lehesjoki, N. E. Stone, V. L. Willour, K. Virtaneva, J. Miao, E. D'Amato, L. Ramirez, M. Faham, M. Koskiniemi, J. Warrington, R. Norio, A. de la Chapelle, D. R. Cox and R. M. Myers. 1996. Mutations in the gene encoding cystatin B in Progressive Myoclonus Epilepsy (EPM1). Science 271:1731-1734.

# GENOME CENTERS: WHAT IS THEIR ROLE?

## Maynard V. Olson

I hope that this meeting will be the start of a broadening discussion of what the Human Genome Project is all about and how it fits into society. I am often thought of as one of the technocrats in this project because my research has focused on methods of analyzing genomes: how does one start by extracting DNA from a sample of human blood and then make one's way from an unpromising glob of translucent material to billions of base pairs of accurate DNA sequence, ready to deposit in a public data base? Viewed narrowly, this task is a problem in bioanalytical chemistry. I was trained as a chemist and tend to think of genome analysis in this way. However, viewed more broadly, the historic movement to analyze completely the human genome needs more than a technical definition: indeed, it poses sweeping challenges to science and society that transcend the project's bioanalytical focus. In this discussion, I want to examine the relationship between the large Genome Centers that are expected to gather the bulk of the Human Genome Project's data and the scientific and social context in which the work will be done.

This visit has provided me with my first view of Tuskegee University. As I walked about the campus yesterday, I was struck both by how much has changed and how much has stayed the same since the University took root here in rural Alabama. At the Carver Museum, I read the inspiring story of Dr. Carver's belief that, in the social context of his day, aggressive application of science was needed to improve the lot of poor southern farmers. The Carver era was a time when there was a simple relationship between a science such as analytical chemistry, at which he excelled, and the society in which it was practiced. Carver's time was also a time of heroes. The South was then dominated by poor farmers whose ability to advance themselves depended on improved yields, new crops, and new markets; progress in understanding the chemistry of peanuts, the microbiology of peanut plants, and the chemistry of the soil in which peanuts were grown translated directly into an improved life for the people with

The author is Professor of Medicine (Division of Medical Genetics) and Genetics at the University of Washington, Seattle, WA 98195.

whom Carver worked. Similarly, in its educational function, Tuskegee University was a symbol of the human ability to respond to a historic challenge—in this case, the sudden need for a university to serve a large segment of the population in the South which had previously been denied access to advanced education. I learned at the museum that Carver himself had wandered about the South and Midwest for more than 10 years in search of a college that would allow him access to the modern knowledge of his day.

The world, of course, has become a more complicated place. Tuskegee University faces a more complex social environment in which to provide education for its students. Similarly, the science that goes on here and elsewhere faces this same trend. This evolution toward an increasingly nuanced relationship between science and society provides the critical background to this meeting, and I welcome the chance to participate. My role is to provide a window into the work that goes on within Human Genome Centers, where much of the analytical work of the Human Genome Project is being carried out. From this story, I will attempt to extract some lessons already learned and look ahead to future trends.

The basic analytical strategy of the Human Genome Project is sometimes called a "divide and conquer" strategy. This approach is required because of the vast amount of information encoded, in linear order, along the DNA molecules in human cells. When viewed in cross section, DNA molecules have atomic dimensions; however, if fully stretched out, they span a length of meters. One can think of the human DNA in a cell as a giant tangle of cooked spaghetti. However, if one kept the proportions right, a strand of spaghetti the length of the human genome would stretch for 1,000 miles. On this scale, to determine the base sequence, a measurement would have to be made every half millimeter, along the strand of spaghetti. Whenever fine measurements must be made across huge distances, the problem must be divided, often successively, into multiple steps (hence, we make maps of the whole country, of individual states, and of particular cities rather than attempting to put street-level detail on a single giant map of the United States).

Recombinant-DNA methods provide the key to the Human Genome Project's divide-and-conquer strategy (see Figure 1). Typically, one starts from a small amount of human blood from which the white blood cells are isolated. Each cell contains two complete copies of the human genome. By starting with a large number of cells, as many copies of the genome as needed are present in the sample. The DNA molecules are broken down at random to produce a huge collection of molecules of appropriate size, which can then be individually propagated as recombinant-DNA molecules in microorganisms such as bacteria or yeast. The advantage of this ability to propagate individual molecules as DNA "clones" is that it is possible to prepare as much sample as needed for the analysis of any particular segment of the human genome.

The process proceeds in successive steps; this lends organization and hierarchy to the analysis of the genome. At the first level, relatively large DNA molecules (typically hundreds of thousands of base pairs) are propagated as

# Divide-and-Conquer Strategy

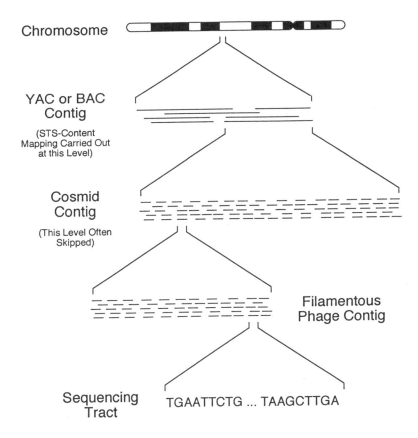

**Fig. 1. Divide-and-Conquer strategy for sequencing the human genome.** The key point in this figure is the concept of successive levels of fragmentation of human DNA into progressively smaller pieces—propagated as recombinant-DNA "clones"—until the pieces are small enough to be suitable for direct sequencing. In the scheme illustrated, only the large-insert YAC or BAC clones (represented as horizontal lines) are prepared directly from human DNA. These clones are built into a "contig" (i.e., a set of overlapping clones that collectively provide redundant coverage of a contiguous segment of the genome). Individual clones from the contig are then chosen and broken into smaller pieces which are used in contig building at the next lower level.

DNA clones. Relatively coarse analysis of these clones allows construction of a rough map of the genome. There are two aspects to this type of mapping. One involves defining a "contig," which is jargon for a densely overlapping set of clones which collectively provide cloned coverage of a contiguous region of the genome. Consider for example the "YAC" (yeast artificial chromosome) or "BAC" (bacterial artificial chromosome) level of Figure 1. In this drawing, a seven-clone contig provides redundant coverage of a targeted region of a human chromosome. This cloned coverage provides essential material to move on to finer levels of analysis.

However, clone contigs provide a poor basis for inter-laboratory cooperation on the sequencing of the human genome, particularly in a project that is expected to extend over a 15-year period. Maps based on contigs alone can only be shared between laboratories by exchanging the actual bacterial or yeast strains that contain the recombinant-DNA molecules. There are frequently problems with stability of the clones during long-term storage. It is awkward for each Genome Center to deal with the many different host strains and recombinant-DNA vectors used at all other Genome Centers. Finally, recombinant-DNA techniques improve steadily; hence, clones prepared in one year are often obsolete in the next. For instance, until the early 1990's, YAC clones dominated standard practice in Genome Centers where the BAC system is now predominant. In a few years, still other systems may offer the best source of cloned coverage of the human genome.

The problem of potential over-dependence on the clone resources associated with particular contig-building projects was largely solved in the late 1980's by adoption of the proposal to base coarse mapping of the human genome on landmarks known as "sequence-tagged sites" or STSs (Olson et al., 1989). Viewed abstractly, an STS is simply a short, defined sequence of human DNA that can serve as a molecular "tag" for a single, specific site within the human genome. Sequences that are repeated more than once within the genome are not suitable for defining an STS. However, it is not difficult to find as many single-copy sequences as necessary to produce a detailed STS map of the genome. In practice, most implementations of the STS concept rely on the polymerase-chain reaction (PCR) to detect specific STSs. A particular clone can be scored as containing or not containing an STS simply by determining whether or not the clone can serve as template for PCR amplification of the STS.

A method known as "STS-content" mapping allows simultaneous construction of clone contigs and STS maps (Figure 2). STS-content mapping simply depends on testing each member of a set of clones that define a contig with a suitably large number of STSs (Green and Olson 1990a). The result of these tests is an "STS-content matrix," which shows which clones contain which STSs. Efficient methods exist to analyze STS content matrices in terms of STS maps. These maps can readily be exchanged between laboratories simply by electronic communication since they depend on only the DNA sequences that define the STSs. Then, when a particular laboratory wants to re-build a contig

# STS-CONTENT MAPPING

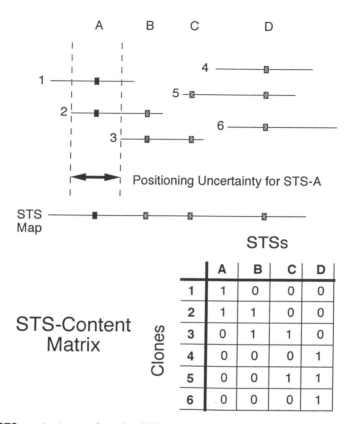

Fig. 2. STS-content mapping. An STS is a "Sequence-Tagged Site," which is simply a short (i.e., typically a few hundred base pairs) stretch of human DNA sequence that "tags" a particular site in the genome. In this drawing the filled boxes are STSs, which have been given the names A, B, C, and D. The STS-content mapping involves testing each member of a set of clones from a "contig" (see Figure 1) for the presence or absence of the STS. In the example shown, six clones—numbered 1-6—have been tested, producing a matrix that shows the presence (matrix element = 1) or absence (matrix element = 0) of each STS in each clone. Analysis of the STS content matrix allows ordering of the STSs in a linear STS map. There is uncertainty in the precise position of an STS mapped in this way: for example, the presence of STS-A in clones 1 and 2, but its absence in clone 3, localizes the STS to the interval shown. In practice, STS-content mapping faces the additional uncertainty that the precise positions of the clone ends are also uncertain. Hence, STS-content mapping provides an excellent means of ordering STSs, but the precise spacing between them is left undefined.

across the region with whatever recombinant-DNA technology is most appropriate at the time, the contig can be built simply by STS-based library screening methods (Green and Olson, 1990b).

The overall divide-and-conquer strategy depends on choosing an appropriate clone at one level (e.g., the YAC or BAC level of Figure 1) and then repeating the whole process of breaking that clone down into smaller pieces that are individually propagated as new recombinant-DNA clones. As the level of analysis becomes more refined, different methods are typically used to organize contigs and develop maps of regions of the genomes. For example, digestion of the clones in a contig with restriction enzymes provides an alternative to STS mapping for determining the overlap relationships between the clones in a contig and abstracting a map from these clones—in this case a restriction-fragment map (Figure 3). More elaborate methods, based on the analysis of multiple restriction digests of each clone, carried out with different restriction enzymes, have also been developed (Wong et al., 1997).

Sometimes there are as many as three successive levels of contig building before clones are produced that are small enough to have ideal properties for direct DNA sequencing. Current methods of DNA sequencing allow 500-1000 base pairs of sequence to be determined by a single analysis on a particular small-insert clone. Given the three-billion-base-pair size of the human genome— and the need to analyze a number of clones from each region to insure high accuracy and complete coverage—the number of small-insert clones that must be analyzed is in the tens of millions. Although there has been great progress in the development of relatively efficient, reliable methods of DNA sequencing, it is still a good day's work for a lab technician to analyze 100 clones. The simple arithmetic of the situation suggests that the data-collection phase of the Human Genome Project will absorb thousands of person-years of effort.

Much of this effort goes into everyday laboratory procedures: growing the microbial cultures that are used to propagate recombinant-DNA molecules, extracting and purifying DNA from these cultures, and carrying out enzyme reactions on the purified DNA in order to gain information about the map position or actual sequence of the particular clone. Electrophoresis, an analytical technique that allows separation of DNA molecules on the basis of the rates at which the negatively charged molecules migrate toward a positive electrode through a hydrated-gel medium, plays a major role in DNA analysis. Technicians in Genome Centers spend a great deal of time casting gels, loading DNA samples onto them, carrying out the electrophoresis, and monitoring the results of the separations.

All this effort has been surprisingly difficult to automate. Excellent instruments are available to carry out the actual electrophoretic analyses—the rate of migration of the DNA through the gel is typically determined by detecting the fluorescence of dye molecules that are attached to the DNA. However, the process of shepherding samples through the process—and, particularly, recognizing when a step has not gone properly and needs to be repeated—is still done most effectively and cheaply by skilled human labor.

Implementation of this process in Genome Centers is still at an early stage. Perhaps 1% of the human genome has been sequenced by the methods described and not all of that sequence yet meets adequate standards of base-pair accuracy and continuity. Much of the large-scale sequencing carried out to date has been applied to the so-called model organisms—yeast, *Drosophila*, the nematode *Caenorhabditis*, and so forth. If one adds up sequencing capacity devoted to these organisms and to the human, existing individual Genome Centers have capacities that span the range 4-40 Mbp/yr. Only the largest center or two are at the upper end of this range, while most of the remaining ones are closer to the lower end. There are only about a dozen Genome Centers in the world with experience even at producing a few megabase-pairs per year of accurate sequence that is continuous over long regions.

## MAPPING CONTIGS USING RESTRICTION DIGESTS

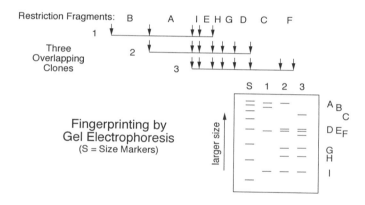

Fig. 3. Mapping contigs using restriction digests. At higher resolution than that typically achieved by STS mapping, restriction-fragment mapping is often used to define contigs and to produce maps of local regions of the genome. The restriction-fragment content of a clone is readily determined by gel electrophoresis, a method that separates restriction fragments by size. Hence, for example, clone 1 is cleaved by a particular restriction enzyme into the four fragments A, B, E, and I, which are separated in lane 2 of the gel shown. A pattern of shared restriction fragments between two clones suggests that the clones overlap. Just as in the case of STS-content mapping (Figure 2), the orders of restriction fragments can be determined from the fragment content of a set of clones in a contig. Sometimes, the local ordering of fragments is not determined by this method: for example, in the drawing shown, the relative order of the clusters (E, I), (D, G, H), and (C, F) are not determined by the data shown.

When carried out on this scale, the cost of DNA sequencing is somewhere in the vicinity of $0.50/bp. Hence, at today's cost, the sequencing of the complete human genome would cost approximately $1.5 billion. Costs will probably drop by roughly a factor of two by the time the bulk of the sequencing is carried out, but I doubt that they will drop much more than that. Thus, the Human Genome Project is on the order of a $1 billion endeavor.

The major issue that presently preoccupies Genome Centers is "scalability." Given the challenges associated with growing to a large scale of operation, it is a plausible assumption that most of the data will be collected in roughly five centers during a period of about five years. This scenario would require that individual centers develop capacities of over 100 megabase-pairs per year. Each current Genome Center wonders which of its processes will break down first as the center attempts to increase its scale of operation by a factor of 10 or more in order to achieve the needed worldwide capacity.

Having sketched out the types of activities that go on within Genome Centers, I would now like to touch on the relationship between these centers and the rest of the scientific community. The centers themselves are largely interdisciplinary units in which the human resources, as is the case in most organizations, are the most critical asset. Most of the instrumentation used in Genome Centers can be purchased commercially or can be adapted relatively easily from commercial instrumentation. As I indicated above, the level of automation in Genome Centers is much less than is commonly represented in press coverage of the Human Genome Project.

Much more important than sample-handling automation is the automation of data processing through effective computer software. In many Genome Centers, as much as half the staff is directly involved in processing the data internally; that is, carrying out all the steps that lie in between obtaining raw output from instruments and the submission of finished maps and sequences to public databases. Of course, there are also a number of technical workers. A typical technician in a Genome Center is a recent college graduate who plans to do technical work for a few years between graduating from college and going on to graduate school, medical school, or other pursuits. I would encourage students from Tuskegee and other institutions to consider this way of spending a few years after they graduate from college. Technical work in a Genome Center only occasionally becomes a career for college graduates. However, it provides an opportunity for a student to immerse himself or herself in a complex technological environment and to learn about technical and scientific careers by observing the large numbers of different role models present in Genome Centers.

I would like to end my presentation with a policy point. Fascinating as the challenge is of building an organization that can sequence human DNA effectively on a large scale, I do not think the concentration of sequencing within Genome Centers is an inherently desirable result. Large centers are certainly required. The only strong justification for Genome Centers is that relatively large organizations—roughly two dozen people is a reasonable estimate—are required to achieve the interdisciplinary critical mass required to carry out all the steps between acquisition of a blood sample and deposition of finished data in a public database. In contrast, economies of scale and the requirement for an elaborate physical facility are relatively minor considerations.

Hence, although I believe we need Genome Centers to pursue the goal of a complete human sequence, I see no strong rationale to concentrate activity in them to the exclusion of other modes of participation in the Human Genome Project. Indeed, I believe that the project itself, and the Genome Centers that are immersed in it, need to have strong interactions with the rest of the biological research community—and with society at large—to prosper.

To policymakers, I would like to propose a new idea that I think deserves careful consideration. As we move to the peak phase of human-genome sequencing, I think that a significant fraction of the actual data collection—perhaps 10 to 20%—should be dispersed outside of Genome Centers even though this commitment would undoubtedly somewhat increase the cost of sequencing the human genome. Such an initiative would greatly broaden awareness within the higher-education system of what the Human Genome Project is all about. It would also provide a powerful means of technology transfer from the Human Genome Project into other worthwhile types of biological research. Modern methods of DNA analysis provide a common underpinning for diverse fields of basic biology, as well as for much biomedical and agricultural research. Hence, it would be a great advantage for faculty and students at institutions such as Tuskegee University to study and do research in immediate proximity to applications of state-of-the-art methods of DNA analysis.

Indeed, the advantages would extend beyond biologists to other members of the University community—those interested in, or concerned about, the implications of advances in our ability to analyze human DNA. The Human Genome Project needs vigorous debate both about its means and its ends. Nothing would do more to inform this debate than dispersal of its central activity into diverse educational and research environments.

The idea is perfectly practical, although it would be admittedly challenging to implement. The satellite facilities would need to have partnering relationships with Genome Centers since they would lack the full critical mass of interdisciplinary activity required to support all the steps in genome analysis. With such partnering—particularly when it involved some exchange of personnel—satellite facilities could function effectively. Rigorous quality-control procedures will be required throughout the data-acquisition program; application of these procedures to both Genome Centers and satellite facilities would insure a high quality for the final product.

Genome Centers are a real success of modern interdisciplinary science. They demonstrate that universities can put together complex teams of researchers who can produce high-quality biological data of broad utility. However, important as team building and interdisciplinary science are in modern research, other goals and values of the research enterprise also require attention. Three important values would be promoted if the data collection for the Human Genome Project were shared between major Genome Centers and satellite facilities: maximizing the access of researchers with diverse biological interests to state-of-the-art technology; encouraging broad participation of students—graduate students, undergraduates, community college students, even high school students—in advanced research; and insuring that broad societal discussion of the future of genetics is informed by widespread familiarity with the technical activities of the Human Genome Project. These activities are too important to society to be overly concentrated at a few sites in the hands of a small number of expert practitioners. This conference at Tuskegee University is a powerful symbol of the breadth of interest throughout society in the Human Genome Project, and I hope that it can be a step toward broader dissemination of knowledge about the project—and even of the actual sequencing of the human genome—beyond the confines of large Genome Centers.

**Post Script (10/97):** As a direct outgrowth of the Tuskegee Conference on the Human Genome Project, steps are underway to establish a satellite facility at Tuskegee University itself. Advanced instrumentation for carrying out four-color-fluorescence sequencing has been acquired by the University and conference organizer Dr. Ed Smith has spent several months at the University of Washington Genome Center studying the Center's technical operations and participating in its work. Upon his return to Tuskegee for the 1997-1998 academic year, he will begin collecting human DNA sequence at the University, as well as using the satellite facility to further his ongoing projects in agricultural genetics. Plans are under development to insure ongoing support of this satellite facility through personnel exchanges both with the University of Washington Genome Center and with other Genome Centers represented at the Tuskegee Conference.

**References**

Green, E.D., and M.V. Olson. 1990a. Chromosomal region of the cystic fibrosis gene in yeast artificial chromosomes: A model for human genome mapping. Science 250:94-98.

Green, E.D., and M.V. Olson. 1990b. Systematic screening of yeast artificial-chromosome libraries by use of the polymerase chain reaction. Proc. Natl. Acad. Sci. USA 87:1213-1217.

Olson, M., L. Hood, C. Cantor, and D. Botstein. 1989. A common language for physical mapping of the human genome. Science 245:1434-1435.

Wong, G.K.-S., Yu, J., Thayer, E.C., and Olson, M.V. 1997. Multiple-Complete-Digest (MCD) Restriction-Fragment Mapping: Generating Sequence-Ready Maps for Large-Scale DNA Sequencing. Proc. Natl. Acad. Sci. USA 94:5225-5230.

# AFTER THE END OF THE BEGINNING:
## THE STORY OF THE BREAST CANCER GENE

Marilyn E. Thompson

The race to find *BRCA1* began with the publishing of a paper entitled "Linkage of Early Onset Familial Breast Cancer to Chromosome 17q21" (*Science* 250:1684-1689, 1990). The senior author of this paper was Dr. Mary-Claire King at the University of California-Berkeley School of Public Health. For years, Dr. King had been theorizing that there was a gene that, when altered, resulted in an increased incidence of breast cancer at an early age. Affected women were diagnosed with breast cancer before menopause and sometimes as young as their 20's and 30's. The diseased gene could be inherited from either parent. By using specific markers, the King lab localized this gene to the long arm of chromosome 17, a very important finding in what was to follow as a competitive race. This gene was named *BRCA1* to indicate breast cancer gene 1 because at the time it was believed that there would be other genes also responsible for breast cancer. We now know that to be true as *BRCA2* has been found and localized to chromosome 13.

Dr. King's work was corroborated in 1991 with the publication of "Familial breast-ovarian cancer locus on chromosome 17q12-q23" by S.A. Narod et al. (*The Lancet* 338:8283). The authors of this paper typed individuals of five large families with increased incidence of breast and ovarian cancer at the same locus used previously by King. They, too, found linkage with breast cancer. Additionally, they also found linkage with ovarian cancer, suggesting that mutations within *BRCA1* also confer increased risk of developing ovarian cancer.

Another key event in the race to find *BRCA1* also occurred in 1991. A biotechnology company was formed in Salt Lake City by Mark Skolnick and Walter Gilbert. One of the primary aims of this company was to find *BRCA1*. However, this was not the only group putting major effort into finding this gene. In addition to the King lab at Berkeley and the newly founded company, Myriad Genetics, Inc., other competitors in this race worked in Bethesda, Ann Arbor, Montreal, London, Cambridge, Glasgow, Tokyo and other places.

The author is currently a NSF postdoctoral fellow at the Vanderbilt University Cancer Center, 649 Medical Research Building II, Nashville, TN 37232.

Over the next three years, progress was steadily being made, using markers specific to chromosome 17 and systematically narrowing the region in which *BRCA1* was localized. Several families were studied by group—those individuals with breast and/or ovarian cancer and the nondiseased individuals. Included in these families were at least two African-American families—one studied at the University of Michigan and another at Myriad Genetics in Utah. It was in the summer of 1994 that, in the African-American family studied in Utah, a mutation in a gene in the vicinity of where *BRCA1* was thought to be was found. This mutation was present in the family members with the disease but not in the healthy individuals. This group proceeded to characterize five families with mutations in this area and published their results in *Science* 266:66-71, 1994. The paper,"A Strong Candidate for the Breast and Ovarian Cancer Susceptibility Gene *BRCA1*" was authored by Miki et al. with the senior author being Mark Skolnick. After four years, the race had been won.

Based on the findings in the Miki et al. paper, *BRCA1* was characterized as encoding a protein of 1,863 amino acids. There is a region at the amino terminus that is consistent with being a zinc finger of the C3HC4 motif. It has 22 coding exons over 100 kb genomic DNA.

*BRCA1* is mutated in approximately 45% of hereditary breast cancer cases and at least 80% of families with increased incidence of breast and ovarian cancer. Although no mutation has been found in any truly sporadic case of breast cancer, they have been identified in approximately 5% of sporadic ovarian cancers. Most of the mutations are frameshifts or nonsense and result in the premature termination of translation. There have been some missense mutations characterized, including those in the zinc finger of the protein. It appears that there are a large number of polymorphisms, so that most of the variation in *BRCA1* is not disease related. Also, the age of onset and the degree of risk are independent of the mutation site. The probability of an affected individual or family having a mutation in *BRCA1* is dependent on the age of diagnosis as well as the number of family members with breast or ovarian cancer.

One question that still remains is: what is the function of this gene product. We had observed that *BRCA1* mRNA levels were higher in normal mammary epithelial cells than in cancer cells and questioned if *BRCA1* expression affected cell growth. To answer that question, we used antisense oligonucleotides to inhibit the expression of *BRCA1*. MCF-7 cells—a human breast cancer cell line—were treated with the oligonucleotides for up to four days. The cells were then harvested and counted. MCF-7 cells grew at a faster rate in the presence of the oligonucleotides. This was a dose dependent effect. Similar results were seen in human mammary epithelial cells, but not in retinal pigmented epithelial cells, a nonmammary cell line. These data suggested that *BRCA1* did have a role in cell growth.

To investigate this more thoroughly, we transfected *BRCA1* into a variety of cell lines and monitored their growth. The cells tested were fibroblasts, human breast cancer cell lines, human ovarian cancer cell lines, colon and lung cancer cell lines. Not only was wildtype *BRCA1* transfected into the cells but also several mutants, including a mutant that terminated the protein at amino acid 1835, a mutant that terminated the protein at amino acid 340 and two large internal deletion mutants. Wildtype *BRCA1* almost completely inhibited the growth of breast and ovarian cancer cell lines, while not affecting the growth of any of the other cell lines. The 1835 Stop mutation inhibited the growth of the ovarian cell line but not the breast cancer line or any other cell line. None of the other mutants affected the growth of any of the cell lines examined. When nude mice were given estrogen pellets and injected with MCF-7 cells, tumors developed. When the MCF-7 cells had been transfected with the retroviral vector, 5/6 of the animals developed tumors within 4 weeks. By 6 weeks all six of the mice had tumors with an average weight of 590 mg. If the mice received cells that had been transfected with *BRCA1*, no tumors were present after 4 weeks. Although by 6 weeks 4/6 of the mice had tumors, the average weight was only 60 mg.

The effect of *BRCA1* on pre-established tumors was also assessed. Tumors were established in animals by the administration of estrogen pellets and nontransfected MCF-7 cells. After tumors had developed in all the animals, the mice were injected with either wildtype *BRCA1* or a nonfunctional mutant. Within 11 days post injection, all the mice receiving the mutant *BRCA1* were dead. The tumor weight at death ranged from 650 to 840 mg. The mice that received wildtype *BRCA1* lived up to 41 days post injection with the largest tumor size being 460 mg. These data indicate (1) that the retroviral transfer of *BRCA1* gene inhibits the growth of breast and ovarian cancer cell lines and (2) wildtype *BRCA1* inhibits the development of tumors in nude mice and suppresses the growth of established tumors.

The biochemical properties of *BRCA1* include a pI of 5.18, having 16% of its amino acids as glutamic or aspartic acids, having multiple dibasic and multibasic sites which are potential proteolytic cleavage sites and numerous potential phosphorylation sites. It also has a 10 amino acid motif that is homologous to a sequence that is found in the granin proteins. Its significance is yet to be understood. In addition, similar to the granins, *BRCA1* is localized to the Golgi apparatus.

The data presented here suggest that *BRCA1* is a Golgi associated protein that functions as a growth inhibitor and tumor suppressor. Since it does not contain a signal peptide, the mechanism by which it is localized to the Golgi requires further investigation. Likewise, the molecular bases for its tumor inhibition ability also requires further studies.

## References

Holt, J.T. et al. 1996. Growth retardation and tumour inhibition by *BRCA1*. Nature Genetics 12: 298-302.

Jensen, R.A. et al. 1996. *BRCA1* is secreted and exhibits properties of a granin. Nature Genetics 12:303-308.

# HYPERTENSION IN THE AFRICAN DIASPORA: GENES AND THE ENVIRONMENT

## Richard S. Cooper, M.D.

### Introduction

In this presentation I will move down the continuum from single gene disorders —like cystic fibrosis—and those caused by relatively uncommon genes— like *BRCA1*—to the complex disorders that affect many more people, exemplified by hypertension. I am going to address the potential power of molecular techniques to help us dissect the genetic basis of complex disorders, but at the same time highlight what I think are some of the limitations of these methods. These limitations include the potential to stigmatize populations, much as we have stigmatized individuals with specific mutations. Throughout I will focus quite narrowly on the issue of hypertension in blacks and how we confront the ideological and social constructs that are used to construct our understanding of disease. The essential message, therefore, will include an optimism regarding the enormous potential becoming available to biomedical researchers as a result of the Human Genome Project. At the same time, however, I will suggest that effective use of these new tools to solve old problems will not be possible until there is an escape from the ideological constraints which have been around for as long as science itself.

Hypertension is a common disorder in all the populations of the world and occurs in about 70% of black Americans by the age of 60 (1). In the rural South in the 1960s it accounted for up to 40% of deaths in blacks (2). For the population of the United States as a whole, hypertension accounts for about 20% of adult deaths among blacks, compared to 10% among whites (3). We have been confronted with this conundrum of the black:white differential for as long as there has been epidemiological research on blood pressure. Not only do blacks have higher rates of hypertension than whites, they also have higher rates than

The author is Professor and Chair, Department of Preventive Medicine and Epidemiology, Loyola University Stritch School of Medicine, Chicago. He is also principal investigator of the International Collaborative Study of Hypertension in Blacks based at 2160 South First Ave., Maywood, IL 60153.

Mexican-Americans (3). The rates are two-fold higher if a more stringent definition is used (i.e., a blood pressure > 160/95 or treatment), but about 50% higher with respect to the common definition (blood pressure ≥ 140/90 or treatment). An understanding of the causes of hypertension within populations, and what might lead to the difference among populations, is urgently needed. A focus on the "ethnic paradigm," as this approach is sometimes called, is also relevant because it demonstrates some of the key weaknesses in the strategy of epidemiology.

**Black-White Differences in Hypertension**
The focus on contrasts in hypertension among blacks and whites in the U.S. has obscured larger differences among black populations. Figure 1 shows the distribution of blood pressure using published data from Nigeria, Barbados and the United States. Blood pressure rises substantially crossing from Africa to the Caribbean into the U.S. Although we do not have detailed data, these populations have a high degree of genetic resemblance. The blood pressure trends are therefore not primarily a result of differences in allele frequencies. It is a public health truism that at the population level it is most often the social environment in which people live that accounts for the mean levels of a risk factor, while among individuals genetic predisposition may play a larger role.

The preceding figure summarizes the level of epidemiological science when we initiated the International Collaborative Study on Hypertension in Blacks in 1991. We were interested, in the broadest sense, in tackling the problem posed in the 1930s by historians who re-examined the relationship between blacks in

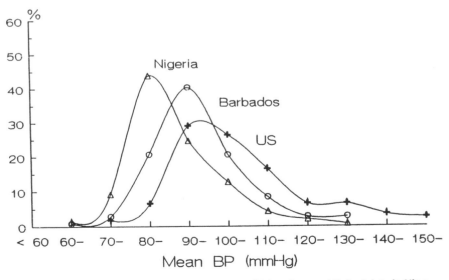

Fig. 1. Frequency distribution of blood pressure (Taken from published data in Nigeria, Barbados and the United States).

the Western Hemisphere and Africa in a modern context. Melville Herskovitz, one of the best known of these historians, helped develop a new view of black history. He argued that the "myth of the Negro past" was debilitating in part because it did not allow us to make the necessary connections and have the social context in which we understood the black experience in this country. He put it this way in his 1941 book:

> The myth of the Negro past is one of the principal supports of race prejudice in this country...it rationalizes discrimination...influences the shaping of policy...and affects the trends of research.... Where all its elements are not accepted, no conflict ensues—even when certain tenets run contrary to some component parts—since its acceptance is so little subject to question that contradictions are not likely to be scrutinized too closely....
> The system is thus to be regarded as mythological in the technical sense of the term, for it provides the sanction for deep-seated belief. (4)

I am going to use his last phrase—"a mythological belief in a technical sense, for it provides the sanctions for deep-seated belief"—as the precedent for the assertion that we have many mythical beliefs about the biology of black Americans. In particular, we have many ideas in public health that have been promulgated over the years, taking on new form, but staying the same in content. This ideological framework likewise has made it very difficult for us to address contemporary health challenges presented by multifactorial diseases. Blacks in the U.S. have higher mortality from virtually all major common diseases (Table 1). Is it reasonable to expect that genetic predisposition is so widespread in this group?

**Table 1. Frequency of Common Health Conditions in Blacks, Whites and Hispanics, US, 1993.**

| Health Condition | Blacks | Whites | Hispanics |
|---|---|---|---|
| | Rate Ratio[1] | | |
| **Prevalence:** | | | |
| Hypertension | 1.5 | 1.0 | 1.0 |
| Obesity | 1.2 | 1.0 | 1.3 |
| Non-Insulin Dependent Diabetes Mellitus | 2.0 | 1.0 | 2.0 |
| **Mortality:** | | | |
| Coronary Heart Disease | 1.1 | 1.0 | 1.0 |
| Stroke | 2.0 | 1.0 | 0.9 |
| Breast Cancer | 1.2 | 1.0 | 0.8 |
| Lung Cancer | 1.4 | 1.0 | 0.5 |
| Prostate Cancer | 2.1 | 1.0 | 0.8 |
| Infant Mortality | 2.0 | 1.0 | 1.0 |

[1] All conditions compared to whites.

At the heart of the myth of genetic predisposition is the belief in the intrinsic inferiority of people of color. Many investigators, for example, believe that there is an intrinsic difference in the impact of blood pressure elevation in blacks compared to whites. A 1994 article states: "Irrespective of the origin of the higher blood pressure, blacks tend to experience greater cardiovascular and renal damage at any level of pressure" (5). I regard this as simply another formulation of the idea of intrinsic inferiority: no matter what the reason for the increase in blood pressure, at the same level blacks demonstrate greater cardiovascular and renal damage—there must be some intrinsic quality about this group that challenges the integrity of their vascular system. In fact, the epidemiological data clearly show that this statement is incorrect (6), reinforcing Herskovitz's dictum that the belief derives from a mythological system. At the heart of the argument for "inherent differences" is the belief in genetic determinism.

Not only are racial differences subject to genetic explanations, there is a tendency in medical biology in general to exaggerate genetic information as it tells us about complex phenotypes. For example, Ganten and Lindpaintner (1991) claim that:

> All basic information on cellular function is stored in the genes....
> The complete set of DNA, the genetic information inherited by an organism is called genotype. Its expression determines the appearance and function of the organism, its physical form, the phenotype. (7)

Clearly something is missing in this formulation—the environment in which the organism develops. There are any number of examples that have already been given in this symposium to show that genes are only expressed in a specific environment and do not mean anything in isolation. The simplified view of genetic determinism is very convenient, however. If one assumes that genes determine the phenotype, it becomes a short logical leap to say that since blood pressure is higher in black populations this is likely to be genetically determined. In its general form the argument goes as follows: since genes determine the phenotype, if a trait is inherited, the difference between groups must be genetic. This construct, known as the "heritability hangup," is dismissed by most geneticists (8) although its influence in biomedical research persists. For example, Law et al. adopt this logic when they write: "Blood pressure seems to be higher among African-American, Caribbean and other black populations at the same level of sodium intake...(and this) difference is likely to be genetically determined." (9)

**Health in the African Diaspora**

It was in an attempt to bring a different perspective to bear on this problem that we began our study (10). Our goal was to examine hypertension in blacks free of the difficulties one has when the outcome is a comparison between blacks

| | CROSS-SECTIONAL SURVEYS | FAMILIES | COHORTS |
|---|---|---|---|
| Africa | Nigeria<br>   Rural<br>   Urban<br>Cameroon<br>   Rural<br>   Urban | Nigeria | Nigeria |
| Caribbean | Jamaica<br>St. Lucia<br>Barbados | Jamaica | Jamaica |
| USA | Chicago (Maywood) | Chicago | Chicago |

Fig. 2. Population samples for the study of chronic disease in the African diaspora.

and whites in the U.S. That comparison is confounded by many factors, the most important being the social experience of people of African descent, and there is no coherent theoretical framework for examining differences. To conduct this study, we chose population samples in West Africa—Nigeria and the Cameroons, both urban and rural, and three islands in the Caribbean—Jamaica, St. Lucia and Barbados—and the working class community of Maywood to the west of Chicago (Figure 2). With this design we have a gradient of risk with differing environmental exposures, providing a laboratory to examine how hypertension is expressed. Hopefully, we can thus avoid both myth-making and the problem of confounding that occurs in the comparison to white Americans.

For the "source population" we went to the Yoruba-speaking region of southwest Nigeria. In collaboration with investigators from the University of Ibadan, we chose the village of Igobrra-pa. This region is just north of the city of Abeokuta, the home of Wole Soyinka, Nobel Prize winner in literature. Long before Soyinka, this has been a society of great cultural accomplishments. The populations from this region were involved in the historic migration to the East during the African iron age and were dispersed from Kenya down into Zimbabwe. In contemporary society there is great diversity in Nigeria as a whole, and many languages are spoken within a short geographic distance.

A typical participant recruited into our study had very low cardiovascular risk. On average, cholesterol was 130 mg/dl, and systolic blood pressure around 110 mmHg. The local subsistence agriculture requires high levels of physical activity. The diet contains only 18% of calories from fat and about 110 milliequivalents of sodium. While the probability of a heart attack in this community is close to zero, parasites are common and, of course, malaria takes a huge toll among the young. The average annual mortality in this village is about 2% among adults compared to about .9% in the United States. This is the environment, therefore, which is kindest to the human cardiovascular system, but harshest to the immune system.

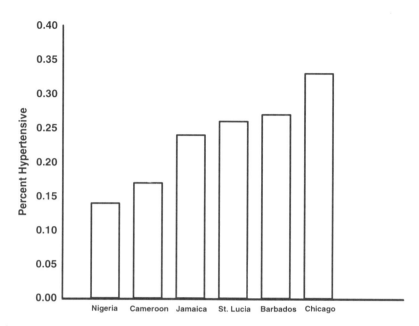

[1]Hypertension defined as Systolic Blood Pressure ≥ 140 or Diastolic Blood Pressure ≥ 90 or taking anti-hypertension medication.

**Fig. 3. Prevalence of hypertension[1] among six populations of West African Origin: The ICSHIB Study, 1995.**

Figure 3 presents the results from surveys which included about 1600 people in each of the six sites (10). A very sharp gradient of hypertension risk exists from 12% to 15% in rural Africa to about 35% in Chicago. It is against this environmental gradient that we can now ask questions about mechanisms (Figure 4). In particular we want to know the role genes play. Are they modified by environment? Can we define gene-environment interactions?

Our first step was to examine the known environmental risk factors. The best described factor is obesity, defined by the body mass index (weight/height$^2$)—a measure of how round a person is. In Figure 5 you can see a straight linear relationship between hypertension and obesity across populations. The meaning of obesity in a mechanistic sense at the individual level is not clear, however, and it is likely to be a proxy for low physical activity, high food intake and a variety of other things. For the present, it is the best physical marker that we have of the risk of hypertension and serves as a useful marker of a lifestyle that predisposes to hypertension.

Obesity makes a substantial contribution to the cross-population differences. Based on relative risk, which is the risk among those who are overweight compared to the risk among those who are not, hypertension is increased by about

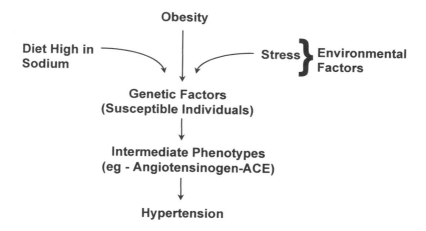

Fig. 4. Pathogenesis of hypertension

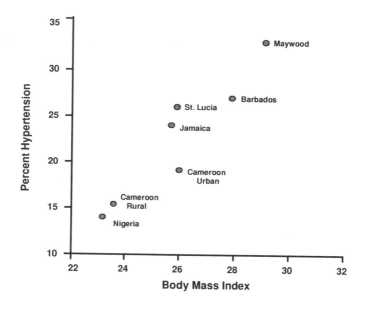

Fig. 5. Prevalence of hypertension by mean body mass index among seven populations of West African Origin: The ICSHIB Study, 1995.

50%. This finding is observed in all our populations (Figure 6). What is also interesting is that the relative risk stabilizes at about 1.5 for all these populations. Thus, even in rural Africa—where people are generally very lean—one

**Hypertension**
**Odds Ratios (95% ÇI) and Population Attributable Risk**
**Comparing Persons with BMI < 25 to those with BMI ≥ 25**

| SITE | ODDS RATIO | PAR % |
|------|-----------|-------|
| Nigeria | 1.8 (0.3--9.7) | 13.0 |
| Jamaica | 1.5 (0.9--2.6) | 20.0 |
| St. Lucia | 2.5 (1.4--4.4) | 43.8 |
| Barbados | 1.6 (0.9--2.8) | 28.3 |
| UK | 1.6 (0.8--3.2) | 28.8 |
| US | 1.8 (1.2--2.7) | 35.0 |

[Adjusted for age and sex]

**Figure 6**

finds the same expression of this risk factor. This finding was somewhat unexpected but says that this factor is generally independent of the degree of obesity and a mixture of other factors. This example therefore suggests one way in which genes can work—the same genes could express themselves in the same way in very different environments.

The next challenge presented is one of understanding the gene-environment interactions (11). Agricultural studies routinely show the effect of different environmental settings on different strains of a single plant species. For example, some corn will do better than others if you modify the environment. Altitude, rainfall and other conditions will mean as much as the adaptablity of the strains. Across a range of environmental exposure, given a specific genetic background, there will be a range of phenotypes. Groups at different points on the scale of environmental exposure, and different genotypes, may express very different phenotypes given these interactions. Not only will the mean values for a trait change, but the variance will change as well. We thus should have the expectation that there will be an intermediate step in all studies which attempt to model the role of genes and environment related to complex disorders—understanding interactions. Only by looking at genes and the environment in combination will we achieve a coherent understanding.

To examine genetic effects we obviously have to examine related people. In the second phase of our study we began to recruit families for research on the familial aggregation of blood pressure. We have begun to collect data from families in Chicago, Ibadan, Nigeria, and Spanish Town, Jamaica, to look at heritability within each population. Figure 7 illustrates our sampling scheme. We identify probands from the low end of the distribution and the high end and enroll all their first degree relatives to obtain a contrast in genetic susceptibility.

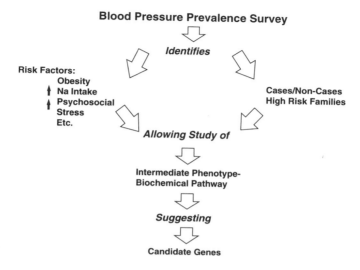

Fig. 7. Current strategy for etiologic research in hypertension.

Genetic material is collected on each of the family members and examined for loci associated with the renin-angiotensin system (RAS). From physiology we know this system is an important regulator of blood pressure, and we have the expectation that variations in the genes that code for proteins in that system might influence the risk of hypertension. In Figure 8 a schematic representation of the RAS is presented. Angiotensinogen (AGT) is a protein made by the liver which circulates in excess and is cleaved by the angiotensin converting enzyme (ACE) to yield the vasoactive peptide angiotensin II. The question for epidemi-

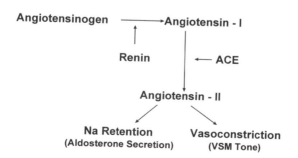

Fig. 8. The renin-angiotensin-aldosterone system.

39

ologists is: Are there relationships between the genes involved in this system, circulating angiotensinogen, the converting enzyme, and blood pressure?

The important finding in this area and, for the moment probably the only published finding of note on genetic factors that influence blood pressure, involves a mutation in the angiotensinogen gene. A "mutant" allele which leads to substitution of tyrosine for methionine at position 235 was found to be more common in hypertensives compared to a control group (12). In addition, this genetic variant is associated with a higher level of expressed gene product. The physiologic explanation is as follows: given that more gene product is being produced, and assuming that it is being cleaved at the same rate, more active product will be available and an increase of high blood pressure would result.

We and others began to look at the frequency of this particular variant in black Americans, and it turns out that about 85% of black Americans have the 235T variant, which is associated with increased risk in whites (13). Among Nigerians, 92% have the T variant. There was then an immediate rush to judgment, if you will, to suggest that the genetic variant that was associated with hypertension in whites is the reason why blacks have more hypertension. Among blacks, however, there is no evidence as yet that this variant has the same association with hypertension risk (13).

In general, I expect this pattern to be repeated. As we begin to dissect out certain genetic traits, there will be an immediate reflex to want to attribute group differences to small, poorly understood genetic factors. This is a process I call stigmatizing populations. We have a long way to go before we can say that angiotensinogen is why blacks have more hypertension. What has been offered is a very quick and simple answer to a very complex problem, and the other concern is that it forecloses our interest and in a way cuts off opportunities to look for the other factors which might cause hypertension among blacks.

Empirical data on the genetics of hypertension in blacks is beginning to emerge. Table 2 presents findings on angiotensinogen gene variants and the level of the gene product in Nigeria and Jamaica. The study in Nigeria demonstrates that hypertensives do have higher levels of the substrate in this physiologic system (Table 3) (14). We have also shown that the individuals that have the 235T allele do produce more angiotensinogen. This relationship does appear to be widespread, therefore, since it works among Europeans and Africans.

**Table 2. Angiotensinogen level by genotype in Nigeria and Jamaica.**

|  |  |  | Genotype* | | |
|---|---|---|---|---|---|
|  | Total | (N) | TT | MT | MM |
| Nigeria | 1539 | (254) | 1561 | 1366 | -- |
| Jamaica | 1682 | (449) | 1625 | 1644 | 1609 |

*position 235

40

Table 3. AGT genotype, level and hypertension, Nigeria.

| Variable | Normotensives (N=138) | Hypertensives (N=116) |
|---|---|---|
| BMI | 22 | 24* |
| SBP | 118 | 162* |
| AGT | 1474 | 1604* |
| 235T | 90 | 92 |

*p < .05
Forrester, Am. J. Hypertension, 1996

Table 4. ACE level by genotype, Nigeria, Jamaica and the United States.

| | Total | (N) | ACE Genotype | | |
|---|---|---|---|---|---|
| | | | II | ID | DD |
| Nigeria | 624 | (1007) | 589 | 653 | 727 |
| Jamaica | 517 | (500) | 423 | 510 | 584 |
| United States | 522 | (107) | 282 | 445 | 552 |

However, the genetic variant itself has not yet been linked directly to hypertension (13).

The other gene I want to touch on briefly is *ACE*. At this locus there is an *alu* motif which results in a common insertion polymorphism associated with variations in levels of activity of this enzyme. Persons with the D allele ('deletion' - or absent *alu* genotype) have higher levels of enzyme activity. We now have evidence that, again, this polymorphism works exactly the same way in persons of African origin as in Europeans. *ACE* is not a rate-limiting step in this system, however, so it is difficult to argue on a physiological basis that it would lead to a higher risk of hypertension. It has been shown to be linked in some studies to risk of other forms of cardiovascular disease, including myocardial infarction and left ventricular hypertrophy. Table 4 contains data from Nigeria, Jamaica, and the US, showing similar relationships among the gene variant and the level of the gene product. On-going work will define the potential relationship to blood pressure.

**Conclusions**

Let me circle back then to where I started in the beginning. We know that there are a number of risk factors which are related to hypertension, and we know that there are differences between blacks and whites in the level of these risk factors. On average, blacks are somewhat more overweight than whites. Sodium intake is really quite similar in the two groups. However, blacks on average consume less potassium, and low potassium is thought to increase the risk of hypertension. We also know that there is this complex set of other factors which are collectively thought of as socioeconomic status and psycho-social stress which undoubtedly plays a role in the risk of hypertension. We thus have a reasonably long list of risk factors which are not well characterized that could also account for the blood pressure difference between blacks and whites.

The technical problem which arises is how to compare the two population groups. First, we have to assume that we have measured these factors accurately in both groups. Next, we have to assume that the measurement tells us the same thing in these two groups. In fact, since neither of those conditions can be met in this case, the whole argument founders. It can be shown as well that the framework of inquiry in epidemiology makes it theoretically impossible to carry out meaningful cross-population adjustment.

As most people in this room are quite well aware, the social experience of being black in the United States is fundamentally and immeasurably different from the social experience of being white. We cannot simply add up the number of years of education, or income, and expect to account for the difference in that experience. Unless we are able to acknowledge that historical fact, we will never be able to use the biological information derived from the "ethnic paradigm." If I may use the analogy, black-white comparisons are like adjusting the number of pregnancies to account for hypertension risk between men and women. In epidemiology we call this the problem of zero exposure. If racism is a causal exposure, we cannot adjust for it. It cannot be measured in both groups and used as the basis for statistical adjustment.

Unless we are prepared to accept the limitation in conceptualizing social exposures, we cannot expect to add genetic factors to the equation and solve the conundrum. I think the temptation to make statistical adjustments for different exposures is the fundamental flaw in the epidemiologic strategy. Over-interpretation may represent as well the chink in the scientific armor which is at great risk of being pierced by the application of genetics to complex disorders. We need to come out of the laboratory and look at the daily news and ask ourselves about the implications of what we are doing. There is now sufficient experience in one area—namely, hypertension—to know that the practitioners of science are going to interpret genetic differences between groups as the result of inherent differences—an outcome we have expected all along.

In conclusion, racial bias not only stigmatizes individuals and destroys the lives of millions of people, it presents a fundamental obstacle to a scientific understanding of human biology. We need to realize that race is not a thing; race is a social idea (Figure 9). When we measure and study it as a data item, it is not only ambiguous, it is deceiving. It is multidimensional and deeply confounded by our constructs of socioeconomic status and culture. At every level, non-comparability of standard socioeconomic measures can be identified. Blacks with a college education make 60% less than whites. Thus, if we adjust for college education, we do not adjust for income. Blacks with the same income as whites pay more for housing, more for car purchases, more for foods and other

**Table 5. The Problem of Racial Bias**

- Race is not a thing, no matter how much we might want it to be.
- As a data item, race is:
    - —deceiving, ambiguous and multi-dimensional
    - —deeply confounded by socio-economic status and culture
- The momentum of historical prejudice:
    - —leads scientists to expect strange and unusual outcomes in non-European ethnic groups ('otherness')
    - —suggests too easily the explanation of intrinsic inferiority ('genetics')

stuffs, so their purchasing power in the marketplace is less. Blacks who purchase a house have a smaller increase in the value of that house over 10 years, so they have less accumulated wealth to then pass on. We cannot actually adjust for inequality in social resources in a society that has systematic racial bias.

The problem, of course, is the historical momentum of prejudice. It is foolish to expect that any of us can individually escape completely and it is complete nonsense to believe that collectively, as scientists, we can escape. The traditions of European thought lead scientists to expect strange and unusual outcomes in non-European ethnic groups—and to suggest too easily the explanation of intrinsic inferiority which has its roots in genes. No fundamental advances will be made in the understanding of ethnic and racial differences in disease patterns until we have thrown off the chains of that racist tradition.

## References

1. Burt, V. L., P. Whelton, E. J. Roccella, C. Brown, J. A. Cutler, M. Higgins, M. J. Horan, and D. Labarthe. 1995. Prevalence of hypertension in the U.S. adult population. Results from the Third National Health and Nutrition Examination Survey, 1988-1991. Hypertension 25:305-313.
2. Cooper, R. S., and C. Rotimi. 1994. Hypertension in populations of West African origin: Is there a genetic predisposition? J Hypertens. 12:215-227.
3. Cooper, R. S. 1997. Hypertension in African Americans. Am. J. Hypertens. (In press).
4. Herskovitz, M. 1941. The Myth of the Negro Past. 1st Edition. Beacon Press, Boston.
5. Kaplan, N.M. 1994. Ethnic aspects of hypertension. Lancet 344:450-452.
6. Liao, Y., R. S. Cooper, and C. Rotimi. 1996. Is hypertension more severe in blacks, or is severe hypertension more common? Ann. Epidemiol. 6:173-180.
7. Ganten, D., and K. Lindpaintner. 1991. Genetic basis of hypertension. Hypertension 18:S101-S109.
8. Feldman, M. W., and R. C. Lewontin. 1975. The heritability hangup. Science 190:1163-1168.

9. Law, M. R., C. D. Frost, and N. J. Wald. 1991. By how much does dietary salt reduction lower blood pressure? 1. Analysis of observational data among populations. BMJ 302:811-815.

10. Cooper, R., C. Rotimi, S. Ataman, D. McGee, B. Osotimehin, S. Kadiri, W. Muna, S. Kingue, H. Fraser, T. Forrester, F. Bennett, and R. Wilks. 1997. Hypertension prevalence in seven populations of West African origin. Am. J. Public Health 87:160-168.

11. Lewontin, R. 1995. Human Diversity. New York: Scientific American Library.

12. Jeunemaitre, X., F. Soubrier, Y. V. Kotelevtsev, R. P. Lifton, C. S. Williams, A. Charru. 1992. Molecular basis of human hypertension: role of angiotensinogen. Cell 71:169-182.

13. Rotimi, C., L. Morrison, R. Cooper, C. Oyejide, E. Effiong, M. Ladipo, B. Osotimehin, and R. Ward. 1994. The role of the angiotensinogen gene in human hypertension: Lack of an association of the M235T allele among African Americans. Hypertension 24:591-594.

14. Rotimi, C., R. Cooper, O. Ogunbiyi, L. Morrison, M. Ladipo, D. Tewksbury, and R. Ward. 1997. Hypertension, serum angiotensinogen and molecular variants of the angiotensinogen gene among Nigerians. Circulation 95:2348-50.

# TECHNOLOGY TRANSFER

**Ronald King**

The tenets of technology transfer basically are very old. In fact, they go back to early civilizations. For example, after the Greeks went to Mesopotamia to get their education in higher mathematics, science and medicine, they returned to Greece and passed that education on to other scholars and also new techniques they learned that represented the technology of that age. That was technology transfer. It may not have been called that, but that is what it was.

If we go to Africa, we can again talk about technology transfer. A form of it could be found in the slave trade. When those unfortunate individuals were packed into those ships and brought to the Americas and the Caribbean, they not only came as beasts of burden or laborers, they also brought with them certain intellectual properties. Farming techniques developed in Africa were applied here. That is a form of technology transfer.

Let us say a word about what technology transfer is not. It is not moving computer equipment from one place to another! Technology transfer is basically a way of transferring the developments and the technology produced through scientific research programs such as the Human Genome Project to the public who will hopefully benefit from that technology.

The National Institutes of Health speaks about its mission when it calls itself "Medicine for the Public." In our Technology Transfer Office at the National Human Genome Research Institute (formerly called the National Center for Human Genome Research), we have the responsibility of transferring gene-based technology into the private sector for further commercialization and development. Everyone has heard of "R & D"—Research and Development. Well, many institutions like the National Institutes of Health and most universities are very big on the "R" but oftentimes much smaller on the "D." There is not much of a commercialization effort. The NIH does not commercialize their products so there is a big jump, for example, from the discovery of the *BRCA1* gene at the laboratory bench and some clinical diagnostic tool for *BRCA1* in the clinic.

The author is chief of the Technology Transfer Office at the National Human Genome Research Institute at the National Institutes of Health, Bethesda, MD 20892.

A lot has to happen for the discovery at the bench to then move out to the pharmacy, clinic or hospital in which it will be used.

I need to give you a little history at this point in speaking about technology transfer. There are two pieces of legislation that come into play: one was in 1980, the Bayh-Dole Act which enacted legislation that allowed intellectual property or the development of government-funded technology to be transferred to the public. In lay terms, let us set up an example to illustrate what took place.

Imagine there is a company on the West coast that makes steel hip-joints for hip-joint replacements, and there is a Department of Defense contractor on the East coast that is developing strong and durable alloys for fighter plane wings. The problem is that the hip joint that is being used in the replacements only lasts about five years. Then you have to go back and do the replacement all over again. So they are looking for more durable alloys that they can use in this hip joint. The problem prior to this legislation was that the technology developed for the airplane was not available to the company needing the stronger hip-joints. The premise was that the taxpayers paid for the development of this research by this company and, if they are not still using it, it is the property of the U.S. government, and other private companies should not benefit from taxpayers paying for the development of this technology. Gradually, this began to be viewed as a short-sighted way of thinking. After the enactment of the Bayh-Dole Act in 1980 that allowed this information to be shared, what would happen would be that the hip-joint company would enter into an agreement with the defense contractor via the federal government, and the information and technology would be shared. The company would then develop hip joints that would have a life span of probably 20 years as opposed to five years. They would sell more hip joints, the company would grow, hire more people, and they would pay more taxes. These taxes would go back to the government and fund more research for more alloys and on and on. So that is what was behind technology transfer and the legislation that came into play.

The second piece of legislation is the Federal Technology Transfer Act that allows federal agencies to enter into collaboration with the private sector, for example, universities or industries, and allows the pooling of resources. As you know, there is only so much money to go around. The Human Genome Project is a very expensive venture and there are a lot of great projects, but not all of them are going to be funded.

There is also a lot of information out there. Some of it is shared and some of it is not. So what you may have is that, as in the hip joint/fighter plane example, there may be a government agency or laboratory with one piece of the puzzle and a private company with another piece of the puzzle. The ideal thing would then be to bring these two together and make it possible to develop a new technology and advance it forward. This is basically what we do. One of the things that we try to do in the Office of Technology Transfer is to try and foster these collaborations and try to speed up the development of technology and try to get it out to the public who can benefit from it.

Now I will fast forward and skip a few things. The Technology Transfer Office is part of the NCHGR intramural program. We have our own intramural laboratories at NIH that are participating in genomic research programs. In addition to our internal research program, we participate in CRADAs (Collaborative Research and Development Agreements), which are collaborations with private industry. In comparing our Institute with the 16 Institutes, Centers and Divisions that make up the National Institutes of Health, we compare very favorably with them when it comes to collaborations. Using 1996 data as an example, NINDS, the National Institute for Neurological Disorders and Stroke, has been in existence for 30 years and is a little over three times the size of the NCHGR; its intramural program had 13 CRADAs. Our intramural program in NCHGR had 11 and was only established in 1993. This gives an indication of just how active our Center has been as far as collaborative efforts go. Although our staff is much smaller in number, we are really big on technology and technology transfer.

One of the threads that runs through technology transfer is a term that has been used several times and that is "intellectual property." The question is what happens when collaborators get together and discover an interesting gene that might be useful as a diagnostic tool. What do you do about this? Typically, one of our responsibilities is securing intellectual property rights for the developments that come out of the intramural program, and these are patent rights.

There is a spectrum of people with various kinds of expertise and so I will give a layman's version of some of the deliberations that private industry, the public and the federal government may have when they deal with things like intellectual property rights. The definition of a patent is "a right granted by the U.S. government to an inventor to exclude others from making, using or selling their claimed invention." It is kind of like the deed to your house. Here is my layman's version:

Let's say there is a very good farmer who has put some money away and has now decided that he wants to go out on his own, get some property and start to grow his own crops. So he searches and finds the ideal piece of property. Let's say it is 200 acres. However, he is only interested in 100 acres.

But he finds another farmer that he sees on the property and says to him, "Do you know who owns the land? I am interested in planting some crops here for this year." The other farmer says, "I own half of this land, I own 100 acres; as far as I know, no one owns the other 100 acres." He points to the 100 acres which looks like it has not been used in quite a while. He says, "I can lease to you my 100 acres of land at a very fair price," and he tells him what the price is and, sure enough, it is a good price.

Then he says, "Or you can go ahead and plant on the other 100 acres that no one owns. But there is a caveat: If you plant on the other 100 acres, there is a chance that someone else might have a slightly larger number of mules or has a larger number of people working for him. He may come when you are halfway ready to harvest your crop and take the crop, and you will have no legal recourse.

That is the risk you take if you plant the 100 acres no one owns." The farmer takes the lease, and that explains why the other 100 acres is not used.

This is something that businesses have to wrestle with. Pharmaceutical companies, for example, have to invest tens of millions of dollars to bring a product to market. If they do not have some type of security, some type of intellectual property right on the pharmaceuticals being developed, they are very reluctant to go into a venture with the risk that someone else might be a little quicker, sharper and take their investment away from them right at the last moment, and they have no legal recourse. If products that come out could not be protected, companies would not be interested in developing them. At least from the industrial side, securing of intellectual property rights is important to the mission of a company and also to the federal government, and it is a way of protecting an investment and an interest.

One example of our collaborations is with Affymetrix which is working with *BRCA1* using the chip technology as a way of trying to identify and detect the many mutations found in a very large gene. We are also collaborating with Research Genetics in using microray technology to look at differential gene expression in up to 20,000 genes simultaneously.

Let me conclude by saying these are some of the really interesting projects that would not be going on if we did not have the type of legislation that would allow us to bring parties together so that they could collaborate and further the technology.

# 2

**Matters of Race and Diversity**

# RACE DIFFERENCES: GENETIC EVIDENCE

## Luigi Luca Cavalli-Sforza

**Summary**

The Human Genome Diversity Project (HGDP) is complementary to the Human Genome Project, adding to it the study of individual variation at the world level. So far, the analysis of individual variation was conducted as part of the discipline called Human Population Genetics and has given many important conclusions, which have already been confirmed and will certainly be extended by the novel techniques allowing the study of DNA directly. Conclusions on races reached so far, and why they debunk racism, are discussed as well as the present state of the HGDP.

The Human Genome Project has set about to study the genome of a single individual, a gigantic task considering the great number of elements (called the "nucleotides") forming its DNA: three billion. Unfortunately, however, this target does not include a severe complication: roughly one of every thousand of these nucleotides is different in pairs of different individuals taken at random. Part of this variation is relevant to medicine, as it causes the difference between health and disease. However, the majority of these differences are, fortunately, trivial from the point of view of social or medical consequences, but are of great interest for other reasons, among them, as markers of human history. The need for studying the individual variation of the genome was stressed in 1990 (L.L. Cavalli-Sforza) and again in 1991 with a group of distinguished scientists. The international arm of the Human Genome Project, HUGO, generated in 1991 a Committee for the Study of Human Genome Diversity of which I was asked to be chairman.

The study of individual variation is not new; on the contrary, it has a long history and is the target of a discipline called human population genetics. Its beginning can be traced to a paper published in 1919 by two immunologists, L.

The author is a Professor of Genetics at Stanford University, Stanford, CA 94305-5120 and is well known for his pioneering work in human population genetics.

Hirszfeld and H. Hirszfeld, who described the variation in world populations of the first human gene ever discovered, responsible for the ABO system of blood groups. The further development of human genetics was slow and difficult; it was marred by sloppy, crude investigations of individual differences for morphological traits, like shape of the nose, color and shape of the eyes and hair, etc. These traits are not well understood even today, but those low quality investigations generated the impression among geneticists of plants and other animals that it was impossible to study human genetics.

There were, however, perfectly sound discoveries of new blood group systems; that of Rh (1940) was very important also from the clinical point of view. When a person is said to be O positive or B negative, or whatever, the first part refers to the *ABO* gene and the second the *Rh* gene (positive or negative). There are other genes that have been discovered since, which are important from a medical point of view, but they were not very many until recently. Until 1980 we knew about 250 to 300 genes using the old techniques available before techniques of direct access to DNA were invented. There are a number among these old "findings" which indicated clinically important variation between individuals. I will remind you only of sickle cell anemia, which allows resistance to malaria, but also causes a serious genetic disease. There are many other genes known today that contribute resistance to malaria. Some of them generate medical problems because some of their carriers may be disadvantaged.

One most important gene (really a family of super-genes) is *HLA*. *HLA* is important in organ transplants and also for the diagnosis and understanding of a variety of diseases. *HLA* was first described around 1960 and has generated an extremely fertile field of genetics. But, as I said, while the genes I named are important for medical reasons, for the majority of others no advantages or disadvantages affecting people who differ for them were discovered. This is one of the many conclusions of interest of human population genetics. There is much variation among individuals—most of it trivial from a social or medical point of view.

I will summarize some other related conclusions to show to people who are afraid of genetics, and particularly of the study of human genome diversity, that long experience has shown that it is a perfectly tame discipline, unlikely to do any harm. In fact, it has been useful, I think, in many ways.

One major aim of the Human Genome Diversity Project is to study human population genetics in an efficient way. The current data we have are almost chaotic. Some studied a few genes in one population; others, a few others in another population. Rarely was it exactly the same genes. Today, we have much more powerful methods and should use them efficiently. The Human Genome Diversity Project is thus the natural complement of the Human Genome Project.

Let me tell you a few of the conclusions that have been reached by straight human population genetics, long before the DNA era. That should reassure those of you who are worried about genetics. But let me say how pleased I am to see in this environment that there is not as much fear as sometimes one is unhappy to discover. I am glad of it; but this is, of course, a selected environment.

One important conclusion of human population genetics is that races do not exist. There is such a remarkable continuity in the variation from place to place that it is practically impossible to define races, except in very approximate ways. If you wanted to be precise, you would have to define many thousands of races and that is not useful. In fact, probably every town could be shown to be different from every other town, but in such a small way, that it really does not matter. But you can detect such differences only if you study a great number of genes. Considering just one gene gives a picture practically incommensurate with that which you would get with other genes.

This was clear already to Charles Darwin. He said, in fact, that it was impossible to define races. Then, as now, some people would see two or three races, or five or six; others would say there are dozens or hundreds of them. This uncertainty is still true today. I do not think there is a way out of it even using the most sophisticated statistical approaches. In fact, such approaches tell us there is no real hope to do a satisfactory classification. But what one can do is try to understand from these data something about the history of humans, and this is one project of the Human Genome Diversity Project.

In addition, it is clear that even if most genes are not important from a medical point of view, there are many that we now know that are. So whatever information we can get at the level of human genome diversity can be important for medicine. It is true, though, that the majority of traits that we know of today—those that are known to be genetically determined—are not important socially. All the attempts that have been made so far to find the genetic basis for socially important traits like, for instance, mental disease have not completely failed, but have been shown to be so complicated that it is very difficult from the analyses done so far to understand the genetic basis of, for example, schizophrenia or other types of psychosis. There are attempts, but these attempts are still uncertain, and conclusions are different so that we are still far from having a clear picture. There is something genetic in much mental disease, but we do not yet know what it is.

Genetic variation is certainly important also for traits like hypertension, described by Dr. Richard Cooper, but also here there is much influence of environmental factors. There are two levels of complication which are extremely serious, where the knowledge from the human genome may contribute—but it will take time. One of them is the interaction between genes, which is often complicated. The other is even more difficult: the interaction between genes and the environment. Part of the problem is that environments, especially those of single individuals, are difficult to measure.

The classic example for that is IQ. I have been interested in IQ because some time ago I realized that a clear mistake was being made in the analysis of IQ data. It had clear social repercussions and was a cause of racism. The mistake is that scientists were forgetting a type of inheritance in humans which is quite independent from that of genes, but it looks almost the same. The results are difficult to distinguish. We call it cultural transmission (or inheritance), using "culture" to mean something very general, which includes the effects of direct or indirect teaching in the family, schools, of social stratification, and also wealth. One should not forget that not only genes, but also wealth and culture are transmitted from parents to children.

It is very difficult to disentangle the two types of transmission, genetic and cultural. Anyhow, a friend and collaborator at Stanford, Professor Marc Feldman, and I wrote a paper published in the *American Journal of Human Genetics* in 1973. The paper was ignored by the people who had been doing studies on heritability of IQ, because they knew nothing about cultural transmission and were boasting that genetic differences were responsible almost entirely for all differences of IQ among individuals. This is what was meant by their estimates of very high heritabilities of 80 or 90%, which we thought were wrong. Other scientists came to a conclusion similar to ours, and estimated that genetic and cultural transmission had about equal importance in determining IQ. One group in Hawaii led by an excellent geneticist, Newton Morton, who was the president of the last International Genetics Congress, also misunderstood our work, at first, and calculated cultural transmission to have practically zero importance. But then other scientists gave results incompatible with Morton's. He repeated his calculations and realized his error, coming eventually to the same conclusion: that cultural and genetic heritability have about the same importance in determining IQ.

The same error about IQ was repeated in the recent book that we all heard about: *The Bell Curve*. This book was published shortly before the translation of a popular book on human evolution I originally wrote in Italian. It was recently translated into English and is called *The Great Human Diasporas*, published by Addison-Wesley. So I had time to add a postscript to the English edition in which I criticized *The Bell Curve* and gave some of the information on the work on IQ which I just told you. The authors of *The Bell Curve* ignored it completely. Whether they ignored it because they did not know about it or because it did not agree with their ideas, I do not know. But you might be interested in reading this postscript to my book and, hopefully, the rest of it.

If races do not exist, one should recognize two things. First of all, racism does exist. What is racism? Racism is defined by many, including me, as the persuasion that some races or racial groups are inherently—that means biologically or genetically—superior to others. What does that mean: superior? Well, all the things racists are interested in are behavioral and socially important traits, but those are the traits that are most difficult to study genetically. Secondly, in general, there is confusion with a totally different social trait and that is *power*, which has very little to do with IQ. Clearly, that is where the confusion lies.

But one should not minimize the idea that there are differences among people, because this is not true. If you look around, you will see that there are differences. I see some black faces, some white faces. I see differences in skin, hair and eye color. Why is there that differentiation? What does it really mean? It took me some time to understand it. I am sure many of you know the answer already. Anyway, let me tell you what my answer is. All these traits that we see are all entirely a question of adaptation to different climates. That includes skin color. It includes body shape. It includes shape of the face and the skull, and so on. In fact, there is no proof to the contrary. The differences of the traits we see are a result of the different climates to which humans have adapted. If you take measurements of all these external phenotypes, you would come up with the conclusion that they have a correlation of 90% with one environmental measure—average temperature. Humidity is also important, but less so.

Is climate very important—especially in an era in which we live in a totally artificial climate? It is clear that it was important in the past. But there has been a lot of cultural adaptation to climate. All these differences that we see in our constitution are mostly due to a small bunch of genes, which are probably not important in other respects. If you want heating or air conditioning of your house to cost less, you have to alter the surface of your house so that there is very little transfer of heat between inside and outside. Well, this is how the body is built. Different bodies have been built in different environments with a surface and a structure that is different in order to protect the individual from the climate to which it is exposed.

Then we have to ask the question: why is it that there are all these differences and we perceive them as so important that we tend to separate humanity according to color of skin and things like that? Why are we so sensitive to this? I believe it is because superficial differences are conspicuous. In other words, they are what we see and we believe what we see. But we merely scratch the surface. We get an impression that there are a lot of differences but, in fact, this is not true, and those that exist are clearly superficial (because they relate to climate adaptation). This is what the data from genetics tell us. There is very little difference among groups. In fact, almost all differences are between individuals within the groups.

There are good statistics on genetic differences among and within populations (Barbujani, 1997). If you take differences between two random individuals of the same population, they are about 85% of the differences you would find if you take two individuals at random from the whole world. This means two things: (1) the differences between individuals are the bulk of the variation; (2) the differences among populations, races, continents, are very small—the latter are only the rest, 15%, about six times less than that between two random individuals of one perhaps very small population (85%). Between you and your town grocer there is on average a variation which is almost as large as that between you and a random individual of the whole world. This person could be from Africa, China, or an Australian aborigine. We will study the differences between two individuals of usually small groups, or as we call them, "populations".

Thus, even if you take a little town, a camp, or a tribe or something like that, you will find an enormous difference among individuals. But you will find only very little differences among them for skin color, because that depends on the particular environment in which the people live or lived. But 85% of all the differences among individuals can be found within populations. What is remarkable is that three calculations based on entirely different types of strictly genetic traits gave almost exactly 85% for the differences among individuals of the same population, even if small.

There is another conclusion that I would like to also cite, because of my special affection for it. If you look at genes, languages, and also at culture, you see a considerable correlation between all of them. The racist conclusion would be: "You see, genes influence culture; genes influence languages." Well, it is rather the other way around. The differences that we have found show that culture (including language) has influenced genes in the sense that different languages and different cultures generate barriers between people and reduce (but only slightly) genetic interchange between them.

One can ask further what are the reasons for this correlation of genes, languages and cultures. Well, the answer is very simple. Much human diversity has been created by the fact that splinters from some groups left the area where they lived and chose to migrate elsewhere to form independent colonies. When they left their place of origin to go to far away places, the mother population and the colonies often evolved totally independently from each other. This was true for both genes and languages. But even if the languages, cultures and genes of the mother population and colonies evolved independently, you still find a correlation between them. Funny enough, Darwin had already reached that conclusion in Chapter 14 of *On the Origin of Species*. But if you are racist and are therefore prejudiced, you are going to misinterpret it and explain every cultural difference as having a genetic origin.

Let me tell you where we are with the Human Genome Diversity Project. It started in 1991, and it has made progress in certain parts of the world. The project has been divided into regions; some regions have been more active than others. The one that has not been active, at least at the level of collecting samples and analyzing them, is North and South America, and Oceania also followed suit. In the rest of the world, things have been going much better. Europe has been fairly active. (The project is not called the HGDP. It is called the Biological History of European Populations, which is what they are currently most interested in studying.) The Chinese government has given money for HGDP research as have India and Pakistan. This makes about one third of the world.

I had a pleasant surprise when I went to Khartoum to collaborate with local scientists. I had been invited there by a Sudanese student who had followed one of our courses and was interested in starting something on his own in his country. I met a wonderful professor of epidemiology who told me: "You know, the Koran promotes the Human Genome Diversity Project." That interested me greatly, of course, and I asked: "How is that?" He said that the Koran says: "I made you men and women of many different nations and tribes, for you to know each other." So I repeated that sentence in the first lecture I gave in Khartoum, and everybody was familiar with it. For the next lecture I learned the Arab equivalent and, when I read it, everybody in the audience chanted it with me because they knew it by heart.

Actually, the Islamic governments have now formed a Union that is going to invest heavily in science and genetics. It so happens that the country which has done the most for the HGDP is Pakistan, where a very good scientist and friend of mine has collected samples from the seven most important populations of Pakistan and has made cell lines of over 100 individuals each. There is no other country that has done as much.

This is good news. There are other places where things are going on. But what about the United States? At the beginning, we had the support of the Department of Energy, the National Science Foundation and the National Institutes of Health for holding an exploratory symposium. However, there were several political problems arising from a campaign by a few political activists best described as a small anti-science organization. This has slowed things down. But in principle, if they are honest, I believe their concerns can be handled. Sometimes you just have to wait.

Two things have happened which I consider excellent signs. One of them is that the National Research Council, which is a branch of the U.S. National Academy of Sciences, has generated a workshop made up of about 15 scientists and ethicists. They are going to examine the feasibility of the project. The workshop is financed by the National Institutes of Health and the National Science Foundation. We will consider the conclusions of this committee with the utmost care. (Their report, made public in the fall of 1997, is clearly favorable to the HGD Project.)

The other excellent news is that the National Science Foundation has put forward a request for grant applications of pilot projects which can essentially examine ethical and practical aspects of the HGDP project. This is the first time there will be money in the U.S. for research on the HGDP, so this is an excellent sign. There are other important signs that things are going well.

I only want to add that I wonder what will happen when the Human Genome Project has finished its task. An army of good scientists and very dedicated young people has been trained and given a lot of specialized equipment. I suspect that some of them, I hope many, will be interested in the Human Genome Diversity Project. If we now start the first obvious stage, collecting population samples—it takes time to get them, then there will soon be a wonderful collection of DNAs so that some of the scientists interested in going from the Human Genome Project to the HGDP will be able to make the best of it.

So I want to say I am happy with the way things are going even if I have been insulted in many different ways by some political activists. A few of them have good intentions, and others just exploited us to promote themselves or their own organizations, using lies of all sorts. But that is, unfortunately, I think, in the nature of human things and we have to accept it.

**References**

Barbujani, G. 1997. An apportionment of human DNA diversity. Proc. Natl.

Acad. Sci. USA 94:4516-4519.

Cavalli-Sforza, L. L., and F. Cavalli-Sforza. 1995. The Great Human Diasporas. Reading, MA: Addison Wesley Publishing Co.

Cavalli-Sforza, L. L. et al. 1991. Call for a worldwide survey of human genetic diversity: a vanishing opportunity for the Human Genome Project. Genomics 11:490-491.

Cavalli-Sforza, L. L. 1990. How can one study individual variation for three billion nucleotides of the human genome? Am. J. Hum. Genet. 46 (4): 649-651.

Cavalli-Sforza, L. L., and M. W. Feldman. 1973. Cultural versus biological inheritance: Phenotypic transmission from parent to child. Am. J. Hum. Genet. 25:618-637.

Cavalli-Sforza, L. L., and W. F. Bodmer. 1971a. The Genetics of Human Populations. New York: W.H. Freeman.

Darwin, C. 1859. On the Origin of Species. London: J. Murray.

Herrnstein, R. J., and C. Murray. 1994. The Bell Curve. New York: The Free Press.

Hirszfeld, L., and H. Hirszfeld. 1919. Essai d'application des methodes au probleme des races. Anthropologie 29:505-537.

Rao, D. C., N. E. Morton, J. M. Lalouel, and R. Lew. 1982. Path analysis under generalized assortative mating II. American IQ. Genetical Research 39: 187-198.

# GLOBAL PERSPECTIVES ON THE HUMAN GENOME PROJECT

## Raymond A. Zilinskas

### Introduction

In a seminal study of international science, Diane Crane deduced that there are four actors that influence the rate at which knowledge pertaining to a new science spreads throughout the world—scientists communicating informally, nongovernmental organizations (NGOs), intergovernmental organizations (IGOs), and national governments (Crane, 1972). The importance of each varies according to the circumstances under which the new science emerges. For example, national governments have played the major role in the international dissemination of nuclear science. Conversely, if we regard genetic engineering, which emerged in the early 1970s, the most crucial roles seem to have been fulfilled by scientists and NGOs (Zilinskas, 1981). Information about the new findings and techniques initially spread throughout the world by scientists communicating informally with one another.

After a few years, scientific NGOs and NGOs whose purview included the biosciences became involved by supporting scientific work, organizing conferences, publishing conference proceedings and reports, and so forth. Soon after that, IGOs began to discern the implications of new scientific developments to their work and sought to determine the steps they needed to take in order to harness this force in the furtherance of their missions.

As to national governments, a few whose scientists were most instrumental in the discoveries that laid the basis for the new science became involved early in the dissemination process, usually by making policy decisions pertaining to funding scientific activities that would propel further developments and regulations to ensure the safety of workers and the public. As the power of the new science became clearer, more and more governments assessed its promises in terms of national economic development and took steps appropriate for national situations. However, there are some governments that even now have made

The author is the Acting Director of the Center for Public Issues in Biotechnology at the University of Maryland Biotechnology Institute, College Park, MD 20740.

little or no attempt to develop national policies that guide their approach toward biotechnology.

Although genetics and genome science are not new, the specific endeavor named the Human Genome Project certainly is, having begun in 1989. It is now approximately seven years later, and the pattern of its global spread appears similar to what occurred in regard to genetic engineering; i.e., first information about scientific developments is spread by scientists communicating informally, followed by the involvement of scientific NGOs, and IGOs whose purview includes related science. Also, similarly to what happened with the world-wide proliferation of genetic engineering, some national governments became participants early in the game, others are now joining in. But most have not demonstrated their interest in HGP developments.

In this paper I describe how an ever-growing number of organizations and countries are becoming involved with HGP activities and analyze why they are doing so.[1] First, I will describe NGOs and IGOs involved in the HGP and explain some of their activities. Second, by considering the countries of Japan, China, and India as case studies, the involvement of governments in the international aspects of the HGP will be reviewed. Third, I will develop ideas on the major implications of the HGP for the enhancement of international science and the development of the biotechnology industry.

**International Organizations and the Human Genome Project**

Both international NGOs and IGOs have become involved with, or indicated their interest in, the HGP. Three NGOs are discussed here—the Human Genome Organization, the Human Genome Diversity Project, and the Human Gene Map Project—and two IGOs: the United Nations Educational, Scientific and Cultural Organization and the World Health Organization. Of course there are big differences between the NGOs, which are dedicated to advancing specialized aspects of human genome research, and the IGOs, which are large organizations whose purview include a broad range of concerns, of which the HGP is only one, and a small one at that.

Human Genome Organization (HUGO)

The imminent commencement in 1988 of the U.S. HGP stimulated scientists in other countries (and the U.S., to be sure) to meet and consider establishing an international organization that would perform similar functions as does the National Human Genome Research Institute in the U.S. The group decided to form HUGO, which was incorporated in England in 1989 as a limited company and registered charity. Full operations commenced after HUGO was awarded significant funding from the English Wellcome Trust in 1991. During 1991 to 1994, English sources, particularly the Wellcome Trust, were the sole sources of HUGO funding. However, beginning in 1994, HUGO has attracted funding from the European Communities, France (Association Francais Contre les Myopathies), Netherlands (Dutch Cancer Society and the Dutch Organization for Scientific Research), and Sweden (Medical Research Council). It continues to receive funding from the Imperial Cancer Research Fund and Wellcome Trust.

According to its statutes, its Board of Trustees decides which activities HUGO is to undertake, while the organization's top policy body is the 18-member HUGO Council, which defines HUGO's scientific agenda and programs (Anonymous, 1993). During its annual meeting in February 1995, the HUGO Council elected Grant R. Sutherland from Australia as the fourth president of HUGO for the term January 1996 to December 1998 (Anonymous, 1995a).

The specific objectives of HUGO are to:
• Help coordinate research on the human genome;
• Foster collaboration between scientists to avoid unnecessary competition or duplication;
• Facilitate the exchange of data and biomaterials relevant to human genome research;
• Promote the spread of relevant technologies;
• Integrate human genome research with parallel studies in model organisms; and
• Encourage public debate and provide information and advice on the scientific, ethical, social and commercial implications of the Human Genome Project (Anonymous, 1993).

Membership in HUGO is open to all persons concerned with scientific subjects related to the human genome. As of September 1, 1996, HUGO had nearly 1,000 members, representing 49 countries. HUGO has three regional offices; HUGO Europe in London, HUGO Americas in Bethesda, Maryland, and HUGO Pacific in Osaka, Japan. Two of the regional offices publish quarterly newsletters; *Genome Digest* in Europe and *Genome Newsletter* in the Pacific.

HUGO brings together representatives from national HGPs and intergovernmental organizations (IGOs) that support these projects. It organized the first International Genome Summit Meeting which was held during January 1994. Delegates from 13 national projects and two IGOs met to identify common areas of interest and define possibilities for international collaborations within these areas (Anonymous, 1994).[2] Discussions at the meeting were held under four headings:
• Databases. There were discussions on the kind and quality of data being stored currently; methods for analyzing stored data; and future needs. Ideas brought up during discussions provided valuable input to the HUGO leadership who will formulate policies that will be recommended to the wider scientific community.
• Ethical, Legal, and Social Issues. Discussions tried to define a coordinating role for HUGO's Ethics Committee on developing "best practices" in the five key areas of genetic counseling, choice of treatments, consent, control of information, and protecting the confidentiality of information.

• Intellectual Property Rights. Delegates unanimously held that full-length or partial sequences of undetermined function should not be patentable.[3]

• Material Sharing. The current arrangement for screening and distributing gene libraries was discussed extensively, as was HUGO's role in coordinating the availability of these resources to scientists throughout the world. The need for better coordination in this area by HUGO was reaffirmed.

The second International Genome Summit Meeting was scheduled for October 1996. A larger number of countries and IGOs participated at this meeting than at the first.

Human Genome Diversity Project (HGDP)

Many of us who are involved in the HGP, even if peripherally, have been asked: whose genome is being mapped and sequenced? The answer that it is a composite human genome, derived from many sources, does not hide the fact that there exist great differences between human populations and between individuals in even a homologous population. For example, if language is used as the criterion, there are 5,000 distinct human populations in the world. Intrigued by these differences, some scientists involved in the HGP began to develop the idea of studying individual genomes (Evans, 1993). Specifically, in 1991 a group of scientists proposed a global project that would aim to collect information about the DNA of populations over the long-term and to set up and maintain an open data base that would contain that information.

Two HUGO-sponsored workshops, held in 1992 and 1993, discussed major issues related to undertaking a coordinated international systematic study of the variations in the human genome. Input from the workshops were employed by HUGO to develop the report "Human Genome Diversity (HGD) Summary Document," which was released at the end of 1994. This report provides a rationale for establishing the HGDP and a format for its organization and work program, as follows:

Many human genes exist in more than one form, and we do not all carry exactly the same forms of every variable gene. Each of us, apart from identical twins, is thus a unique individual, recognizably human but different from all other humans. The genetic variation from one person to another reflects the evolution of our species as it results, over many generations, from the survival or loss of different forms of genes or the natural introduction of new forms. Studying this variation among people from around the world can therefore provide a great deal of information about the development of our species which, integrated with findings from archaeology, linguistics, history and other disciplines, can lead to a much richer and more complete picture of our past than has been previously possible. As a cultural resource, the potential of the project is therefore enormous (Anonymous, 1995b).

The HGDP was developed under the auspices of HUGO's HGD Executive Committee under the chairmanship of Dr. Luca Cavalli-Sforza. The HGDP, which began its operations in 1994, will attempt to study several hundred of the world's populations, selecting those that can be considered broadly representative of them all. Samples of blood, hair, and/or cheek scrapings will be collected from individuals of various populations and the DNA content will be analyzed. Collection centers will be established in various regions to receive and store samples and to establish cell lines. In turn, samples and cell lines will be transferred to HGDP central repositories where they will be maintained as a global resource. In view of the possibility that information derived from HGDP activities could easily be misused, great care has been taken by the HGD Executive Committee to draw up ethical guidelines that will govern the behavior of HGDP participants and use of HGDP information. Nine guidelines have been enunciated that address issues such as respect for individuals and their cultural integrity, protection of test subjects from commercial exploitation, control of information derived from HGDP research, and others (Anonymous, 1995b).

The Human Gene Map Project (HGMP)

In October 1994, the Wellcome Trust sponsored a meeting that brought together scientists interested in sharing information about the genes they map with each other and the scientific community at large. Out of this meeting grew the idea of commencing the HGMP, which would be an international effort to find human genes that are expressed, map their location, and deposit information about them in public data bases. A second meeting, held in January 1995, developed strategies for implementing the HGMP. In particular, it was decided at that time to base the map on expressed segment tags (ESTs), which have been explained as being:

...short, identifying sequences obtained by partially sequencing cDNAs. ESTs are obtained from cDNAs represented in arrayed libraries from various tissues. If suitable primers for an EST are designed, PCR can be used to amplify the corresponding sequence from genomic DNA. The EST is thus converted to an STS [sequence-tagged site] that can be mapped to a genomic location using radiation hybrids (RHs) or genomic clones such as YACs [yeast artificial chromosomes] and BACs [bacterial artificial chromosomes] (Stewart, 1995a).

A third meeting on the HGMP was held in May of 1995. Much progress was reported on sequence mapping. For example, while only 1,000 ESTs had been mapped by the end of 1994, 37,500 were expected to have been mapped by the end of 1995 (Stewart, 1995b). At this rate of discovery, the sequence-level map is likely to be finished this year, thus ending the HGMP.

United Nations Education, Scientific and Cultural Organization (UNESCO)

UNESCO's mission includes promoting education, science and culture, involving all countries in these activities, and protecting the dignity and uniqueness of the human species. To further these objectives, UNESCO became involved in the recombinant DNA controversy at an early stage by sponsoring meetings and conferences at which ethical issues stemming from advances in biotechnology were addressed. UNESCO's involvement in the HGP thus is a natural progression of its intense interest in bioethics, to try to make certain that human beings are protected and the human species as a whole is safeguarded from the misuse of information derived from genome science. UNESCO's multifaceted involvement with the HGP is described in a 1993 report (Lenoir, 1993). It began in 1989 when its Director-General appointed Dr. S. Grisolia from Spain as Chairman of UNESCO's Scientific Co-ordinating Committee for the Human Genome Project. The Committee soon defined a framework for UNESCO actions in reference to the HGP, which included organizing meetings of an interdisciplinary character on subjects of global concern, the providing of fellowships for scientists from developing countries, and establishing the International Bioethics Committee.

*International meetings*

UNESCO sponsored six international meetings pertaining to the HGP in 1990, five in 1991, and four in 1992 and in 1993. Perhaps the most important of these were the First South-North Human Genome Conference, held in Caxambu, Brazil, in 1992 and the Second South-North Human Genome Conference, held in Beijing in 1994. During the First Conference, the so-called "Caxambu Declaration on Patenting of Human DNA Sequences" was issued, which in part read "...we urge that consideration be given to avoiding the patenting of naturally occurring DNA sequences. The protection of intellectual property should, in our opinion, be based on the use of sequences rather than on the sequences themselves." The Second Conference issued no such ringing statement but did provide an update on scientific work to developing country scientists and provided a forum for Chinese scientists to demonstrate their considerable contributions to the HGP (see below) (Anonymous, 1995c).

*International fellowships*

During 1990 to 1994, UNESCO provided more than 40 fellowships to scientists from developing countries to study or train in laboratories performing HGP-related research. Further, in 1993 it entered into an agreement with the Third World Academy of Sciences in Trieste, Italy, to provide short-term UNESCO/TWAS fellowships, but I have no information on how many of these have been awarded.

*International Bioethics Committee*

In 1993, the UNESCO Director-General acted to set up the International Bioethics Committee and invited 40 eminent biological, medical, and social scientists to become its members (Lenoir, 1993). Noelle Lenoir from France was named its chairperson. The Committee, the first of its kind, is primarily a forum for the exchange of ideas and for debate. At each of its sessions one specific theme is addressed, such as genome research, embryology, the neurosciences, gene therapy, and genetic testing. During these sessions, discussions may concern the current state of progress in research, possible applications of research findings, identification of ethical concerns that research and/or applications can engender, and ideas as to how these concerns ought to be addressed. To conclude the section on UNESCO, it has become evident that the organization has defined three ways by which developing countries can participate in a substantial way in the HGP. First, their scientists should pay special attention to genetic traits, including inherited diseases, in native populations. Second, governments should organize scientific work using the best scientists they have for mapping and sequencing some representative sites on the human and other genomes, particularly those that may have particular value to the host country. Third, representatives from developing countries should take part in the moral and ethical debates on genetic technology, including safeguarding the rights of individuals (Anonymous, 1995c).

World Health Organization (WHO)

WHO has not taken a direct part in HGP-related research activities, but has so far limited itself to monitoring progress related to what it considers the important ethical, social, and legal aspects of controlling hereditary diseases (World Health Organization, 1994). WHO has deemed eight aspects in particular as important: (1) newborn screening for treatable conditions (screening should be mandatory and include counseling); (2) predictive testing (which should be used only for medical purposes); (3) detecting susceptibility genes (susceptibility information should be provided in terms of risk factors, probabilities, and percentages); (4) human rights and human duties (with the right of privacy being most important); (5) data protection (mechanisms for protecting medical information need to be strengthened and affirmed); (6) genetic counseling by properly trained personnel must accompany genetic testing; (7) decision-making and society (WHO to take a leadership role in disseminating information that leads to sound decisions); and (8) consensus ethics (international surveys and studies should be done by UNESCO, WHO and others to determine areas in ethics where global consensus exists).

## Genome Programs in Japan, China, and India

While scientists worldwide are either involved in genome science research or are following developments in this field closely, comparatively few governments have considered policies that would define the role of their nation's scientists in the HGP. To illustrate, after having reviewed the literature and interviewed scientists in various countries, I find that, in addition to the U.S., the following 11 countries have considered the HGP on the political level: Australia, China, France, Germany, India, Italy, Japan, Netherlands, Russia, Sweden, and the United Kingdom. As can be expected, the level of involvement by governments varies widely—from making substantial resources available to promoting in-country HGP-related activities to offering encouragements in words only. Since Japan, China, and India represent the scope of governmental involvement in the HGP, they are discussed next in detail.

Japan

Similar to other large industrialized countries, Japan's political structure dealing with science and technology issues is large and complex (Zilinskas, Colwell, Lipton and Hill, 1995). The top policy body is the Council for Science and Technology (CST). It provides advice to the Prime Minister, who makes decisions that are implemented by the executive agencies. In the biomedical field, the most important agencies are the Science and Technology Agency (STA), Ministry of Science and Education (MSE), Ministry of Health and Welfare, Ministry of Agriculture, Forestry, and Fisheries, and Ministry of International Trade and Industry. All these Japanese agencies and ministries have become supportive of Japan's HGP (Sakaki, 1995; Anonymous, 1996; Science and Technology Agency, 1994). However, the first was the Science and Technology Agency.

The STA's first involvement with genomic science was through research. Thus, it funded the pilot study "Research for Genomic Organization" in 1987 at the Institute of Physical and Chemical Research (RIKEN), as well as the follow-up project, "The RIKEN Genome Project in 1988." The major achievement of this project was the determination of the whole *Saccharomyces cerevisiae* chromosome VI. However, on the policy side, STA began considering the implications of the HGP for Japan during 1988 to 1989. During 1989 to 1990, the major decisions were made at the inter-agency level that established the HGP in Japan. Specifically, the elements of a two-phase program were instituted.

During Phase I (1989-1995), four major accomplishments were achieved:

• In 1990, the Human Genome Committee was established at the highest level, under the CST, and a CST member, Dr. Wataru Mori, was named its head. Other members of the Committee are the leaders of the two main HGP projects (see below) and representatives from the agencies named above. In a 1991 report, the Committee set objectives for the HGP in Japan, described steps of all related scientific activities in Japan, and recommended roles for Japanese agencies and ministries in achieving objectives (Council for Science and Technology, Panel on Life Sciences, 1991).

• The Japanese HGP officially started in 1991, when the Ministry of Science and Education provided funding that established the environment for genome science research in Japan. Specifically, two large projects were funded. The first, totaling approximately 3 billion yen ($2.72 million) over five years, was provided to Dr. Ken-ichi Matsubara at Osaka University for the project "Human Genome Analysis," while 2.5 billion yen ($2.5 million) over five years was awarded to Dr. Minoru Kanehisa at Kyoto University to support "Genome Informatics." In addition to funding specific research projects, these grants funded institution building, including the establishment of the Human Genome Center at the University of Tokyo, research laboratories for genome analysis at the Kyoto, Kyushu, and Osaka universities, and the computer network called "Genome Net."

• The Science and Technology Agency contributed funds to international activities, including sharing in the cost for establishing the Genome Database in the United States and setting up a Japan Node at the Japan Information Center of Science and Technology (JICST). The research funded by STA led to the development of a fine physical map of human chromosome 21 and high resolution map of chromosome 3.

• Other institutions established during Phase I include the DNA Database of Japan at the National Institute of Genetics, the Rice Genome Project at the National Institute of Agrobiological Resources, and the HUGO Pacific office at the Human Genome Center in Tokyo.

Phase II began in 1996 and will end in the year 2000. Activities and projects encompassing Phase II are outlined in the "Report for the Promotion of the Second Phase of the HGP in Japan," which was issued by the Human Genome Committee in 1994. Unlike Phase I, which concentrates on institution building, Phase II mostly consists of research to achieve substantial scientific objectives, and agencies are tasked to support projects for that end. Accordingly, through JICST, the Science and Technology Agency is supporting an attempt to sequence the human genome on a megabase scale and, also, a RIKEN project that focuses on the functional analysis of the mouse genome, while MSE is promoting the program "Human Genome Analysis" that will undertake sequence-based analysis of human genome structure, functional analysis of the genome, and bioinformatics pertaining to the genome.

It can be seen that at an early stage of the HGP, the Japanese government decided it would make sufficient funding available to enable Japanese scientists to contribute significantly to that international effort. A fair number of laboratories located throughout Japan are now performing research directly related to the Japanese HGP (Figure 1). The amount of funding that the Japanese govern-

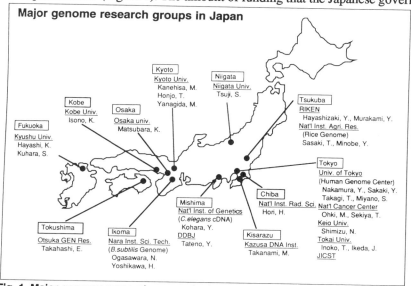

Fig. 1. Major genome research groups in Japan.

ment dedicates to promoting the HGP demonstrates its commitment: approximately 200 billion yen ($182 million) was allocated to fund Phase I activities, and in 1995 alone it spent 51.06 billion yen (or approximately $46.4 million) for this purpose. Although the financial situation in Japan has worsened over the last three years or so, causing cut-backs in government funding in most sectors, the Japanese government has announced that it will increase spending in support of science and technology. Therefore, we can expect that the total support for the Japanese HGP during Phase II will increase by 10 to 15 percent.

China
Similarly to Japan, China is undertaking a phased approach to implementing its HGP. However, China commenced its activities later, in 1994, so the First Phase is scheduled for 1994 to 1997 (Tan, 1996). Plans for the Second Phase have not been published as of this writing.

The now-underway First Phase is sponsored by the National Natural Sciences Foundation of China. Major funding for First Phase projects is provided by the State Commission of Education, Ministry of Public Health, and local governments (the Shanghai Municipality, in particular). The Chinese HGP as a whole is supervised by an Advisory Committee, whose membership is composed of highly respected scientists and government officials, while day-to-day coordination of HGP activities is overseen by an Academic Committee, which is served by a secretariat. The HGP is carried out mainly in five major Chinese cities (Beijing, Changsha, Harbin, Kunming, and Shanghai) at 19 laboratories operated either by local universities or the Chinese Academy of Sciences and the Chinese Academy of Medical Sciences (Figure 2).

The First Phase is constituted by three programs: Resource Conservation, Technology and Informatics Development, and Studies on Disease Genes.

*Resource Conservation*
As a result of its large population (1.2 billion) and population diversity (56 ethnic groups), China provides unsurpassed genetic resources for human genome diversity research. Stimulated by the HGDP, Chinese scientists are trying to capitalize on these resources. Accordingly, blood samples and cheek scrapings are being collected from ethnic groups and cell lines are being established from representative samples. Biodiversity studies are underway using selected loci, including the mitochondrial DNA, and Y chromosomal and other genomic markers. Special attention is being given to large families afflicted with a distinct disease or possessing certain phenotypes.

*Technology and Informatics Development*
In recognition of China's lack of scientific resources, the aim of this program is to modernize data bases, electronic communications, bioscience research facilities, and scientific equipment to enable scientists to fully enter into human genome research (eg., China installed its first 64-kilobit Internet link only in 1994). Accordingly, yeast artificial chromosome and cosmid genomic libraries are being established, techniques for genetic identification such as genotyping

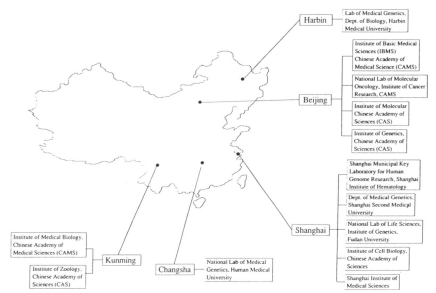

**Fig. 2. Major genome research groups in China**

are being introduced, and an informatics network is being set up that will enable Chinese researchers to communicate with one another and with international information centers. So far, 36 Chinese scientists are HUGO members.

*Studies on Disease Genes*

This project focuses on high-frequency genetic diseases in China and identification of disease genes. Thus, novel genes associated with esophageal and hepatic cancer have been identified, genes associated with exostosis have been cloned, and molecular studies on non-insulin diabetes have begun. In addition, disease loci for Wilson's disease, beta thalassemia, Marfan syndrome, and others are being investigated with the aim of developing gene diagnosis and therapy.

In addition to research, efforts are underway to educate the public, medical officers, and clinicians about the significance of genomic research and the value of medical genetics. For this purpose, education and training programs have been developed for use in the mass communications media, and in lectures and courses in secondary schools and universities.

The Chinese government has not published the amount of funding it is dedicating to national and international endeavors related to the HGP. We can deduce from the foregoing that it is not an insignificant sum, and it is used mostly to address health problems afflicting the Chinese population.

India

Major responsibility for biotechnology in India rests with the Department of Biotechnology (DBT), which is part of the Ministry of Science and Technology (Zilinskas, 1993). As part of its effort to enhance India's capabilities in biotechnology, the DBT is taking steps to expand genetics research.[4] Earlier this year, the DBT set up an advisory committee composed of scientists known for their work in genetics. Its head is Dr. O. Siddiqui, the director of the National Centre for Biological Sciences. Next, the DBT commenced a two-pronged approach in genome science. The first, named "Technology Development and Manpower Training in Human Genome" is headquartered at the Indian Institute of Science in Bangalore. Under this program, the Institute has, or is in the process of completing, four tasks: (1) developed gene mapping software that enables researchers to compare megabaselong DNA sequences; (2) developing *E. coli* and yeast model systems to investigate human genetic disorders; (3) building a tunneling microscope system to image DNA molecules; and (4) cloning and sequencing genes from *Entamoeba histolitica* for use in developing drugs to control amoebic dysentery, a major disease in India.

The second approach, which was stimulated by the establishment of the HGDP, is entitled "Genome Analysis of Genetic Disorders Common in India." As its name makes clear, it addresses genetic diseases, including thalassemia, Duchenne muscular dystrophy, retinitis pigmentosa, and juvenile myoclonic epilepsy. Activities undertaken under this approach will focus on developing molecular approaches to the analysis and management of genetic disorders. This will entail the collecting of samples from indigenous populations and their analysis, research to find and characterize disease genes, and the establishment of a network of genetic clinics throughout the country that will provide genetic counseling to afflicted individuals and families.

Of the three countries discussed in detail here, only India seems to have encountered potentially serious problems related to genetic science research. Specifically, some Indian researchers are concerned about the export of human DNA and blood samples to foreign research groups. They worry that information derived from this material may be developed overseas for commercial purposes without Indian consent and no provision for sharing profits with the Indian population (Jayaraman, 1996). This problem came to a head earlier this year when researchers at the U.S. National Institutes of Health were accused of violating an Indian regulation forbidding the export of biological material without permission from the Indian Council of Medical Research (Jayaraman and Macilwain, 1996). Although this particular episode seems to have been peacefully settled, the problem is likely to reemerge, if in a slightly different form. The reason is that many foreign research groups are working in India, mostly in collaboration with Indian researchers, because India offers promising research opportunities due to its huge population, prevailing social patterns (including large families, consanguineous marriages, and stability of isolated communities), diverse cultures, and a variety of rare genetic mutations. For example, a large population in West Bengal appears immune to cholera. It should be possible to employ a technique called differential genome comparison to identify the gene or genes responsible for this immunity and develop the protein it or they code for. Other similarly promising commercial opportunities may abound.

## Conclusion

Imagine, if you will, trying to explain to a politician why she or he should invest public monies to help fund an international project that is mapping the entire human genome. One approach might be to rationalize that, if scientists were able to catalogue the 100,000 or so human genes, they could apply the genetic code to deduce the sequence of amino acids that constitute the proteins whose composition each of these genes direct. Knowing this, they next could clarify the three-dimensional structures and functions of the proteins. This accomplishment, in turn, would enable scientists to discover how proteins and nucleic acids interact in cells, work out the details of physiological processes, and explain the process of cell differentiation. Once all this had been accomplished, our descendants would be well positioned to apply this knowledge to, for instance, develop methods for preventing or ameliorating disease.

I think that the reader would agree with me that this approach, if not leading to a blank stare, would be unlikely to persuade the politician to divert scarce public funds from, for example, immediate poverty alleviation or improving primary health care delivery.

Perhaps a more compelling approach would be to explain how knowledge of the human genome can be applied for economic development. When we regard the sources of biotechnology (Figure 3), the major source of knowledge and applications today are biotechnology research projects collectively that focus on microorganisms, plants, and animals. It is safe to predict that as the HGP culminates in the first decade of the 21st century, it will have become the second major source of biotechnology. As a consequence, the development of diagnostic and therapeutic pharmaceuticals, as well as gene therapies, will occur in an entrepreneurial environment depicted in Figure 3.

If we regard Figure 3 with its flow of developmental stages from left to right, the input today from the HGP into biotechnology entrepreneurship is on the order of several magnitudes less than from non-human genome projects. What do I mean by this statement? In practical terms: of the number of applications in the pipeline, only a small fraction have their genesis in the HGP. However, in ten years or less, the balance between the two sources will begin to equalize as more and more novel knowledge generated by the HGP is translated into new products and therapies. Who knows, as the 21st century advances, the HGP and its successors might become *the* major source of biotechnology!

The situation as depicted in Figure 3 is no secret; it just has not been well recognized by those who make political decisions that affect the conduct of science and technology in most countries. But this is changing rapidly. Worldwide publicity about intellectual property rights concerning human genes and achievements of biotechnology companies dedicated to applying results from the Human Genome Project has apprised the educated public everywhere about the HGP and its promises (and its possibilities for misapplications, to be sure). As a result, most decision makers by far in countries that possess a pharmaceu-

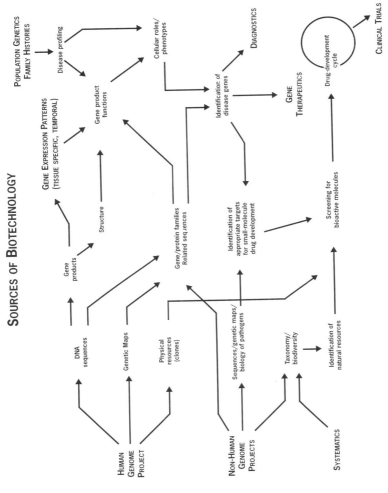

**Figure 3.**

tical industry, whether nascent or mature, are cognizant of the HGP and its objectives. Therefore, an ever increasing number of them are becoming convinced that it is necessary for their nations' scientists to take part in the HGP, so they either are funding activities for that end or are about to do so.

**Endnotes**

1. I do not address the subject of scientists communicating informally about HGP-related activities because no one has, as far as I am aware, studied it formally. However, for the purpose of this paper I assume that a very large number of interactions have taken place among scientists that culminated in the estab-

72

lishment of the HGP in the United States and a few other countries such as Japan and the United Kingdom. I base this assumption on the fact that the governments of nations were able to institute national human genome projects quickly, something that could not have been done unless scientists and their organizations in these countries were able to mount powerful, convincing campaigns to persuade politicians to adopt the laws required to allocate public monies to fund these efforts.

2. Australia, Canada, Commission of the European Union, France, Italy, Japan, Latin American Biotechnology Network, Netherlands, Russia, United Nations Educational, Scientific, and Cultural Organization (UNESCO), United Kingdom, United States, and the World Health Organization (WHO).

3. In early 1995, HUGO released a 15-page statement on patenting DNA sequences, which in part reads "HUGO is worried that the patenting of partial and uncharacterized cDNA sequences will reward those who make routine discoveries but penalize those who determine biological function or application. Such an outcome would impede the development of diagnostics and therapeutics, which is clearly not in the public interest. HUGO is also dedicated to the early release of genome information, thus accelerating widespread investigation of functional aspects of genes. This statement explains our concerns." (Anonymous, 1995a)

4. Unlike Japan and China's efforts in genome science, of which greater (in the case of Japan) or lesser amounts of information have been published, I have found no literature references pertaining to India's activities related to the HGP. Therefore, except as indicated, the information in this section is anecdotal, derived from interviews with Indian scientists and officials.

**References**

Anonymous. 1993. Who is HUGO? Genome Digest 1:2-3.
Anonymous. 1994. HUGO hosts first genome summit. Human Genome News 6:8.
Anonymous. 1995. News from HUGO. Human Genome News 6:5.
Anonymous. 1995b. The Human Genome Diversity (HGD) Project. Genome Digest 2:12-15.
Anonymous. 1995c. Spotlight on China: South-North Conference. Genome Digest 2:8.
Anonymous. 1996. The Human Genome Project in Japan: the second five years. HUGO Genome Digest
Council for Science and Technology, Panel on Life Sciences, Human Genome Committee. 1991. Plans to Promote Work on Analysis of the Human Genome – Report of Human Genome Committee. (Tokyo: Council for Sci-

ence and Technology).

Crane, D. 1972. Invisible Colleges. Chicago: University of Chicago Press.

Evans, L. 1993. Human genome diversity. Genome Digest 1:7.

Jayaraman, K.S. 1996. Indian researchers press for stricter rules to regulate "gene-hunting." Nature 379:381-382.

Jayaraman, K.S. and C. Macilwain. 1996. Scientists challenged over unauthorized export of data. Nature 379:381.

Lenoir, N. 1993. Study Submitted by the Director-General Concerning the Possibility of Drawing Up an International Instrument for the Protection of the Human Genome. UNESCO document # 27C/45, (Paris: UNESCO).

Sakaki, Y. 1995. The Human Genome Project in Japan. Hugo Pacific GENOME Newsletter 2-7.

Science and Technology Agency. 1994. Answers to the questionnaire coordinated by the Science and Technology Agency, Japan. In Proceedings of the International Genome Summit Meeting, held January 20-22, 1994, in Houston, Texas, (Bethesda, MD: The Human Genome Organization).

Stewart, A. 1995a. Human gene map workshop held. Human Genome News 7:7-11.

Stewart, A. 1995b. More on the Human Gene Map initiative. Genome Digest 2:6-9.

Tan, J. 1996. Genome Project in China. Hugo Pacific GENOME Newsletter 1-4.

World Health Organization. 1994. Ethical (E), social (S) and legal (L) aspects of controlling hereditary diseases as a result of progress being made by the Human Genome Project. In Proceedings of the International Genome Summit Meeting, held January 20-22, 1994, in Houston, Texas, (Bethesda, MD: The Human Genome Organization).

Zilinskas, R.A. 1981. Managing the International Consequences of Recombinant DNA Research (Doctoral Dissertation), (Los Angeles, CA: University of Southern California).

Zilinskas, R.A. 1993. Capability-building in biotechnology by developing countries. Lessons from a plant biotechnology project in India. In G.T. Tzotzos (ed.), Biotechnology R&D Trends: Science Policy for Development, Annals of the New York Academy of Sciences, vol. 700, pp. 232-256.

Zilinskas, R. A., R. R. Colwell, D.W. Lipton, and R. T. Hill. 1995. The Global Challenge of Marine Biotechnology: A Status Report on the United States, Japan, Australia and Norway, (College Park, MD: Maryland Sea Grant College).

# THE DILEMMA OF DIFFERENCE

**Patricia King, J.D.**

It was very much like coming home to be able to come to a state and to an institution that played such a role in our own history and to remember that many of the activities started here at Tuskegee University have helped make it possible for me to stand here today. Let me begin by saying that I am not a scientist. I am not a physician. I never majored in science, and I make no pretense of understanding much of what was said by some of the others. Please be assured that you can save your science-based questions for them.

I am here because I am intimately familiar in some ways with the Human Genome Project. I am here to share my concerns and my hopes for this project and the information that it will generate and how we might think about what we should do in the future. We all know that the Human Genome Project is a huge project funded by Congress in the expectation that it would play a profound role in medicine and biology in the next century and that the work that was done would help to alleviate human suffering. The promise is enormous, and I always keep that in front of me.

But I have always thought that the risks to our social fabric of undertaking this project were also great—that from the beginning there was a potential for misuse of the information that was going to be generated. Dot Nelken and Larry Tancredi (1989) said early on in their book: "Testing for the biological origins of disease can affect our concepts of social equality, justice and privacy and our roles about choice and free will..." (p. 167) They continue by stating that "the most important implication of biological testing is the risk of expanding the number of people who simply do not fit.... We risk increasing the number of people defined as unemployable, uneducable or uninsurable. We risk creating a biologic underclass." (pp. 175-176)

Because there was awareness of these risks, something called the ELSI (Ethical, Legal and Social Issues) group was created simultaneously with the Human Genome Project. Along with some others in this room, I was an initial

The author is Carmack Waterhouse Professor of Law, Medicine, Ethics and Public Policy at Georgetown University Law Center, Washington, D.C. 20001. ©May not be reproduced in any form without the expressed written consent of the author.

member of that working group. There are many significant ethical, legal, and social issues raised by the Genome Project. From the beginning, my own personal interest revolved around the concern about whether—and how—African-Americans were going to benefit or be burdened by the information that was going to be generated. I worried and still worry that we are at risk for becoming members of that biological underclass Nelken and Tancredi worried about.

From my perspective, there has always been reason to worry and there are two categories of reasons—or I try to put everything in these two categories—that suggest to me that we have quite a bit of work on our hands. First, the danger is that, for African-Americans and other minorities and the poor, greater attention will be paid to genetically-based explanations than to more complex explanations for society's medical and social problems. This historically has been done and continues in some ways to this day.

The second problem for me has been that the poor and minorities will not even benefit from the improved health outcomes that might be made possible by the genome project. It is not so clear to me—in fact, there is a great deal of evidence to the contrary—that the benefits that we all hope to achieve will be fairly distributed in the society in terms of access to and delivery of health care services.

So I think that a look at the Human Genome Project raises significant concerns for many of us. These are not new concerns. There is nothing special about the Genome Project in terms of it being distinct from other scientific and technological advances that have created problems. But the degree to which the Genome Project raises these issues is of concern. It will result along with other movements in society, I sometimes fear, to our detriment.

Historically, there has been a linkage between genetic information and race. This society has expressed its preference for homogeneity and rationalized differences by appeal to genetic explanations for these differences. I think that there is no escaping that. We will all be misled if we all thought that it would be possible any time soon to disaggregate the society's use of race and its racism from genetic information. Historically, this information has been used to penalize and has served as a method of oppression.

Race was historically appropriated in furtherance of other social goals such as justifying slavery which is really all about how to get free labor. If you have visited Williamsburg and the Williamsburg Foundation Programs—updated to bring them into the 20th century because they did not start this way, you will recall that for a very short time in the 17th century, the legal status of the Negro was not very different from the legal status of the indentured servant. The status of both was relatively unsettled. By the 18th century, however, slavery was entrenched, lifelong and passed to one's children.

How did we go from the relatively same treatment of indentured servants who were white and other servants who were black to eventual freedom for one and slavery for the other? We were able to do that because in some sense the most important and significant trait is passed from parents to children, and it is race. Not biological race. There is no biological significance to race. It is race in the sense that there are visibly apparent differences that others use for their own purposes.

So this link between parent and child was the beginning of the ruling class's ability to control our status and to control what would happen to us. Racial identity was equated and used to justify superiority and inferiority. Whites were born free, blacks were born slaves and one's position in life was set at birth. Racial identity was determined by the mother, not by the father, because we might have a different result if it was determined by the father. It was determined by the mother because the mother in almost all instances was a black woman. We did not have to worry very much about what would happen if a white woman had a baby that was a black man's. Although for some of you who may be interested, Thomas Jefferson himself thought you basically had to get them out of the state. You can read the details of Jefferson's views about what you did in that set of circumstances in the October, 1996 issue of *The Atlantic Monthly*.

The disadvantages and the stereotypes that were spawned in this early connection between what we thought about genetics and what we perceived genetics to be and color of skin have persisted to this day. They have not been unseated by the civil rights movement of the '60s and '70s.

Let me give you another example—one that is often used to show the socially destructive ways that race can be used. Eugenics, as we all know, is the study of the improvement of the human species. It is fundamentally concerned always with reproduction, encouraging those to reproduce who are fit and discouraging or preventing reproduction of those who are not fit. Genetic information was used to support these goals and used in a way to support the view of the inferiority of some minority groups and some social classes.

Eugenics happened a long time ago, some of you may say, and is not possibly of relevance today. I disagree. The ideas that were spawned then play out today in our health care and in our politics. I have so many examples but I will spare you most of them. I will just give you two of the many that I come across in my work. African-American women in recent years have been coerced into accepting sterilization because of threats they would lose their welfare benefits. The link between welfare benefits and sterilization leads directly back to who is fit to reproduce.

In 1990, when the FDA announced that Norplant, a contraceptive, was now available for distribution in the United States, a day later the *Philadelphia Enquirer* editorialized that Norplant ought to be provided to women on welfare and women who use drugs to restrict them from reproducing. The people that they were talking about were poor black women.

A third example (and this is an area where I spend lots of time) is that, as many of you know, there have been attempts at prosecution of pregnant women who use drugs during pregnancy. What some of you may not know is that all the women prosecuted were African-American. You might also reflect on the fact that alcohol, which to date is proven to be much more serious in terms of its impact on the fetus than most drugs and seems to be widely available and used in the population, has not resulted in the same kind of attention. You might ask, why is it that we are distinguishing between women who drink during pregnancy from those who use drugs. I suggest that what we are engaged in is some carry-over and extension of stereotypes and attitudes toward African-Americans, in this case African-American women, that comes from that precise link that is still in the mind between genetics and the racial coloration of people's skin.

I also want to say that medicine and health care institutions have not been immune to some of the same problems. Nor has law for that matter. I do not want you to think that I am here casting stones at other people's direction. With a different audience, I could give a different kind of history, but it would be equally devastating.

Many of us have forgotten, but I have not, that we have in our health care system, vestiges of a segregated, separate health system that we have never really focused on, identified or tried to come to grips with. In my hometown community, the hospital that was the black hospital is still there. It is still a hospital and its presence in the community continues to operate almost exactly the way it did during the years of segregation.

The history of our health system and the history of sickle cell screening and indeed the Tuskegee Syphilis Experiment itself, I think, is testimony to the fact that the health care system itself is not immune from some of the stereotypes and perceptions of African-Americans that have existed. In fact, in the *New England Journal of Medicine* about two weeks ago, there was a study on the effects of race and income on mortality and the use of services among Medicare beneficiaries. The conclusion was that race and income had a substantial effect on mortality and use of services among Medicare beneficiaries. Medicare coverage alone is not sufficient to promote effective patterns of use by all beneficiaries. In the same issue, it was editorialized and I would agree that "race and social class have continued to be powerful and often divisive factors in our national life....they influence the health status, access to health care, and scope and quality of health care of minority and poor populations." These findings are consistent.

Well, where are we? We know what our history is; some of us know it better than others and that may be part of the problem. But we are nonetheless faced, it seems to me, with a great opportunity that comes from the Human Genome Project. The issue is: how can these historical pitfalls be avoided? Indeed for some, the question is: can they be avoided? What can we do?

I think we face an incredible dilemma. The dilemma was first described for me by a professor, a legal scholar, Martha Minow, who has written a great deal about differences. She has emphasized the importance of difference and context in her work. When people have suffered harmful discrimination based on traits such as race, gender, etc., remedies can be structured by rejecting lines of differences and focusing on similarities between people. That certainly is one strategy to take. But she also points out that by ignoring differences other harms may result, unless the differences are recognized in the context in which the differences remain significant.

That for me is the crisis of the Genome Project. That is, where to recognize similarities, how to ensure that African-Americans—it is our tax money too—benefit from this project in ways that others might benefit, but benefit also because in many ways we are different. I am speaking about differences that are socially constructed and which have persisted for generations and cannot be ignored. Information does not fall into a vacuum.

Information in this case is going to fall into a society where racism is pervasive, where ethnic stereotyping and economic inequalities are rampant. Our problem is: how can we talk about our differences and not have our differences used against us—because historically that is how differences have been used.

I think a complicating factor is that, since the Genome Project has been primarily seen as a medical or science-based project, the group to whom the task of analyzing these problems has been given—ELSI, but more broadly bioethics, I think is incapable today of dealing with some of these issues. Why do I say bioethics is incapable of helping us deal with how to structure information or how to control the use of information in ways that do not hurt? Because I think that bioethics in terms of its methodology and its structure is just not set up to help us tackle these problems.

What do I mean by that? Renée Fox, who is an old friend and role model for me and wonderfully insightful, has written, "there is a sense in which bioethics has taken its American societal and cultural attributes for granted, ignoring them in ways that imply that its conception of ethics, its value system, and its mode of reasoning transcend social and cultural particularities." I think she is right. One of the modern emphases in bioethics, for example, is a dependence on principles that were referred to by Elizabeth Thomson when she said that the National Institutes of Health has a project now to examine the values of the Belmont principles in the genome context. The Belmont principles were issued by the National Commission for the Protection of Human Subjects. I was a member of that group too and helped write that report. However, I wonder if the Belmont report is what we need in terms of what to do with genetic information.

I reflect, for example, on how we think about the Tuskegee Syphilis Experiment. In my work, no matter how often we talk about the syphilis experiment, we always come back to the fact that those black men did not give informed consent. They did not, and that is what we talk about. I think the important thing about the experiment is: how did the federal government get away with identifying a county called Macon in Alabama thinking that they could move there and collect a group of men and proceed to do what they did for 40 years. For 40 years! What was behind that? It is not the informed consent; it is a justice issue. It is an issue about how you select subjects. It is a question about the mythology of blacks and whites that led them to Tuskegee in the first place. Syphilis, a sexually transmitted disease, was viewed as being different in African-Americans than in whites. Therefore, the logic went, we need to study some African-American men. The history that I am interested in is not that they did not give full consent. Given the conditions in which these people lived, I think that is a joke to some extent. Unless we can have a bioethics that focuses not only on the informed consent, but asks the kinds of questions about why and what inequities in American society, what inequities in medicine—the systemic aspects—are involved, we are not going to be successful in insuring that the genetic information that is bound to come is not going to be hurtful.

The second way in which I think bioethics is not useful is that it grew out of liberal individualism which says the individual is king. It is a bioethics that focuses on individuals. It does not focus on families. It does not focus on relationships. It does not focus on groups. It does not permit us to talk about what it means to be African-American, or a member of a group. I do not make the mistake in thinking we are all alike, nor do I want to make the mistake of essentialism. But African-Americans as a group have shared experiences that make us all understand that we are African-Americans. You cannot capture those experiences, you cannot capture that knowledge by focusing only on the individual and nothing else. You particularly cannot do it in an abstract theoretical framework.

Where does that leave me? For a long time, in despair. Because as I looked for tools on how to attack some of these problems, I just despaired. But I am optimistic, a little, that it can be done.

I do not have any map or program about how you deal with differences. But I think what we have to do is to begin at the beginning, and the beginning for those of us who are not scientists and not physicians involves understanding, teasing out, going back over 300 years, the ways in which the connection between genetics and race and social good have been entangled. I am most familiar with efforts to do this in the law because I am a lawyer. People like Judge A. Leon Higginbotham, for example, are spending a great deal of time showing how American law, law-by-law, has helped to establish and maintain racism in our society. We need to do it in medicine as well, and we need to do it so that those of us who are outside can continue to speak out against directions that seem to be hurtful.

But those of you who are the scientists and the physicians who actually will help generate the knowledge, you need to do it too. The biggest danger is that a person believes that they have neutral information that will do good. Unless you understand the uses that can be made of your information, you are feeding into the hands of those who would misuse it.

So for all of us, scientist and non-scientist, physician and non-physician, it is important to know the past because that is what we learn from. It is always important to understand as we work that whatever we say will not necessarily be heard the way we say it. You can talk all day about race not having biological significance, but I am convinced that it will be another 100 years—if we are lucky—before Americans will stop acting as though it has biological significance. Until we understand that, my view is that we are nowhere. Next time, I will try to go beyond beginning at the beginning, but I think that is a huge and enormous task.

## References

Renée C. Fox. 1990. The evolution of American bioethics: A sociological perspective. In George Weisz (Ed.), Social Science Perspectives on Medical Ethics. Dorarecht: Kluwer Academic Publishers.

Geiger, H. Jack. 1996. Race and Health Care—An American Dilemma? New England Journal of Medicine 335(11):815-816 (September 12).

Gornick, Marian E., Paul W. Eggers, Thomas W. Reilly, Renee H. Mentneck, Leslye K. Fitterman, Lawrence E. Kucken, Bruce C. Vladeck. 1996. Effects of race and income on mortality and use of services among Medicare beneficiaries. New England Journal of Medicine 335(11):791-799 (September 12).

Higginbotham, A. Leon. 1978. In the Matter of Color: Race and the American Legal Process: The Colonial Period. New York: Oxford University Press.

Higginbotham, A. Leon. 1996. Shades of Freedom. Racial Politics and Presumptions of the American Legal Process. New York: Oxford University Press.

Martha Minow. 1990. Making All the Difference: Inclusion, Exclusion, and American Law. Ithaca: Cornell University Press.

Nelken, Dorothy, and Laurence Tancredi. 1989. Dangerous Diagnostics: The Social Power of Biological Information. Basic Books, New York.

# THE RESPONSIBILITY OF SCIENTISTS IN THE GENETICS AND RACE CONTROVERSIES

## Jon Beckwith

People choose careers in scientific research for a variety of reasons. Some go into science because they want to contribute to the solution of significant social problems, others because they are curious about how nature works, yet others simply because early on they have shown a talent in mathematical and scientific subjects. Many of the daily activities of the academic scientist are in basic research problems that may range in their obvious relevance to social applications from none to very relevant. On the one hand, research areas may start out as obscure, arcane pursuits and then bear fruit as socially applicable findings. On the other hand, scientists who spend their entire careers focusing on some problem of social interest, e.g., cancer, may never see any benefits to society from their efforts. When science does enter the realm of social application, scientists as a community are rarely prepared for the complexity of the science-society interface. There is some truth to the long-standing image of the ivory-tower scientist, pursuing research that may have dangerous consequences without any concern for or awareness of these consequences.

There is an implicit or explicit training of scientists to believe that science is a neutral pursuit and that they are not responsible for its consequences. This training misleads future scientists in two ways. First, personal, social, cultural and ideological biases are an inherent and even essential part of science (Kuhn, 1970; Gould, 1981; Longino, 1990), although we strive to be as aware of them as possible. For example, Sir Cyril Burt studying the "intelligence" of children from the lower social classes let slip his personal disdain for these individuals when he described a child as "a typical slum monkey with the muzzle of a paleface chimpanzee" (Rensberger, 1976). In all likelihood, these prejudices strongly affected his objectivity in evaluating and presenting results and led perhaps to his fabricating results (Kamin, 1974; Hearnshaw, 1979). A possible parallel today can be seen in research into the biological origins of sexual orien-

The author is an American Cancer Society Research Professor in the Department of Microbiology and Molecular Genetics at Harvard University Medical School, Boston, MA 02115.

tation (LeVay, 1991; Bailey and Pillard, 1991; Hamer et al., 1993). Gay researchers have entered this field with the explicit hope that finding genes or biological correlates of male homosexual behavior will make homosexuality socially more acceptable and reduce discrimination. This "hope" perhaps blinds these scientists to the weakness of their arguments and leads to an overreaching in presentation of conclusions to the public. These more obvious examples can be supplemented by numerous other cases to be found in fields of research that are much less socially relevant. In my own field, biases about biological mechanisms being universal led to researchers missing out on the enormous variety of gene regulatory mechanisms during the decade of the 1960's (Beckwith, 1987).

The point here is that assumptions, social or otherwise, underlie any scientific research project and thus color the outcome. This often unappreciated character of science belies the claim for neutrality, particularly in those areas that touch on social concerns.

Second, scientists have a responsibility to inform the public so that it is equipped to evaluate the significance and impact of particular scientific findings. If a scientific finding or claim is clearly going to affect public thinking, social policy or public health, those who have knowledge in the area and see misrepresentations taking place should contribute to the public discussion. While others may occasionally be able to fulfill that role, who better to present and/or debate the scientific issues than scientists themselves? This is not to say that the scientific community will have uniform views on the validity or implications of well-publicized scientific findings. But open discussion by scientists about these issues should, in principle, benefit public discussion and decisions.

This brings us to the interface of genetics and society over the last century. Beginning with the founding of modern genetic research at the turn of the 20th century, geneticists played an important role in applying genetic ideas to human social problems (Ludmerer, 1972; Provine, 1973; Allen, 1975; Kevles, 1985; Muller-Hill, 1988; Proctor, 1988; Beckwith, 1993). The wedding of genetics with eugenics led to a social movement with substantial social impact (Kevles, 1985). Not only was the public led to believe that this new genetics had established a solid scientific basis for racial and ethnic differences in various behaviors and intellectual capacities, but also laws were passed that resulted in discrimination against certain ethnic groups within this country and tremendously reduced immigration from eastern and southern Europe into the United States.

By the time the movement had reached its peak in the 1920's, scientists—including those who had initially supported the thinking and claims of the eugenics movement—were no longer so enamored of it and sometimes privately deplored eugenic activities (Ludmerer, 1972; Allen, 1975). But they failed to speak out publicly on these issues and thus missed an opportunity to raise questions in the public's mind about the scientific validity of the claims. The resulting free scientific hand the eugenicists were given allowed them to move from propaganda and "public education" to influencing social policy. Even worse, in Germany, geneticists continued to support and even lead the way in eugenics thinking and social policy suggestions, providing one of the underpinnings of the Holocaust.

I should make clear that my concerns about potential social consequences of genetics in this country today are not that there will be a repetition of the Holocaust nor even of the kinds of eugenic programs of these earlier periods. Rather, in a society that has been torn during the last 40 years over the civil rights movement, school integration, affirmative action, etc., academic arguments for irremedial genetic differences in capabilities between groups may be used to reinforce and support existing discrimination and to oppose policies that attempt to remedy past injustices.

I should make another point clear. In principle, there is nothing to fear from the truth whether it comes from genetics or some other source. It is the distortion of the meaning of genetics that is to be feared and opposed. A major point I will elaborate on is that those who have used genetic arguments to support the social policies I have alluded to have not only presented highly questionable genetic evidence, they have also misrepresented what it means to find a genetic component of a human trait. What I will argue is that, even when genetic contributors to some capabilities and behaviors are found, they are rarely useful for elaborating social policy.

In the last several years, there has been a revival of racialist science (Mehler, 1994). This revival has been capped off by the publication of *The Bell Curve* by Richard Herrnstein and Charles Murray (1994). While much of the response to this book has been negative, the media coverage of the book has been enormous, and the book has sold close to a million copies. Other indications of a resurgent interest in these arguments are the flap over publication of a book by Christopher Brand which also makes arguments for genetic differences in intelligence between racial groups (Mackintosh, 1996; Bell and Masood, 1996) and a well-publicized speech by a past-president of the Behavior Genetics Association who argued for genetic differences between blacks and whites in empathy, aggression and impulsivity (Butler, 1995). Interestingly, the president-elect of the organization resigned in protest.

Why this current attraction to an argument that has appeared before, been rejected, and still is based mainly on old, questionable studies? It is certainly not because genes have now been discovered that correlate with these behaviors or capabilities. There are no major new studies of other sorts that would explain this resurgent hereditarianism. I would suggest that a major reason is a cultural environment in which the successes of the new molecular genetics have received enormous public attention and in which claims from the practitioners of this science have often emphasized the implications of human genome research for understanding everything about ourselves (Beckwith, 1993). With the spate of publicity for that research in the last few years, one could easily come away with the impression that genes have been found for just about everything and that, as Jim Watson put it, "our fate is in our genes" (Jaroff, 1989).

I have no problems with the media paying attention to these genetic findings. They are exciting and we are learning a lot about human development and human disease. What I do have a problem with is the presentation of misconceptions about the meanings of genetics. Furthermore, the lumping together in terms of scientific acceptability of the precise gene-mapping studies with the older familial studies gives a credibility to these latter studies that is not warranted. While the application of molecular genetics to study certain human behavioral issues promises to put this field on solid scientific ground, previous approaches involving studies of families, of genetically identical twins and adoption studies are subject to numerous complications that render them only suggestive at best (Billings et al., 1992).

The result of the current media barrage could well be a continuation of a major misunderstanding about genetics, that is, the idea that finding a gene associated with some trait or disease means that that trait or disease is fixed—that it cannot be changed by environmental conditions. Furthermore, the continued presentation of new findings as "the gene for X" confuses correlation with cause and effect. For instance, years ago researchers discovered that gene mutations that result in the absence of the metabolic enzyme phenylalanine hydroxylase led to the accumulation of toxic compounds in the brain causing severe mental retardation (Tourian and Sidbury, 1983). A simplistic representation of this finding today might have trumpeted it as "researchers find the gene for intelligence." But, this is a rare mutation and represents only a tiny fraction of the mutations in a variety of other genes that can affect brain function in this way. Furthermore, the condition can be effectively reversed by changing the environment. In this case, altering the diet of children with the condition restores normal brain intellectual function. This example illustrates the need to first understand the biochemical or physiological basis of any genetic correlate of a condition or behavior and the potential for dramatic effects of environmental conditions on the expression of any trait.

Yet, today, the finding of a rare family in Holland where a mutation in another metabolic gene, monoamine oxidase A (Brunner et al., 1993), appears to be correlated with aggressive behavior, is trumpeted by Cowlet and Hall in the November 1, 1993 *Newsweek* with the headline "Genetics of Bad Behavior." Further, the concept of genetic fatalism—that genes fix human traits—is still presented in the media often by scientists themselves. Consider the following quotes:

> • From an article entitled "Some Bed-Wetting Is Linked to a Gene"—Dr. Hans Elberg of the University of Copenhagen states: "The families were very relieved to learn it was genetic, and so beyond the control of parent or child" (NYTimes 7-1-95, p. l).

> • From an article entitled "How Clever is Charles Murray?"—"... he (Murray) claims that intelligence is substantially inherited (so there is

nothing much that traditional social policy can do about it)" (*The Economist* 10-22-94, p. 29).

• From an article entitled "Sex: It's All in Your Brain"—According to author John Leo: "Every social explanation has been exhausted—this is innate. ...is the public ready for the reality that some high-prestige, high-paying fields will be 75 percent male, some 75 percent female?" (*U.S. News & World Report* 2-27-95, p. 22).

• From an article entitled "Born to be Fat"—According to National Cancer Institute researcher Dr. Dean Hamer: "On the one hand, having a gene is practically useful because you can argue that it's an immutable trait" (*The Washington Post*, section on Health, 12-6-94, p. 12).

• From an article entitled "Do Males Have a Math Gene?"—"Benbow and Stanley contend that scientists should first determine the source of sex differences in math ability. If the differences are environmental, they may be able to be eliminated; if they are genetic, we must learn to accept them" (*Newsweek* 12-15-80, p. 73).

These articles alternately quote scientists or their interpreters assuming that genetics means "fixed and unchangeable." These statements have obvious social implications. If girls cannot do mathematics well and it is due to a genetic deficit, then why should teachers and parents encourage girls to study math seriously? If one is "born to be fat," why eat a healthy, lowfat diet? If IQ differences between groups are genetic and fixed, why waste society's resources on trying to educate those with the genetically based lower intelligence.

Ironically, while genetics is presented to the public in this simplistic way, knowledge of human genetics is expanding with the use of new technologies revealing an ever-deepening complexity to ways in which genes are expressed (Abbott, 1996). Consider the following:

First, well before the advent of the new DNA technologies, it had been clear that for a number of genetic conditions there was substantial variation in expression of symptoms (Maddox, 1993; Strohman, 1993; Wolf, 1995). For example, children born with Down's Syndrome vary enormously in their degree of retardation and their physical health. Some die at a very early age, while others go on to lead socially productive lives, well-integrated into society. A recent example is the French actor Pascal Duquenne, who with his co-star Daniel Auteuil from the film "Eighth Day" won a joint Best Actor Award at the 1996 Cannes Film Festival where Duquenne gave an impressive acceptance speech.

But now, with the ability to detect anybody who carries a mutation known to be associated with a particular condition, a number of surprises are emerging. Consider Gaucher's Disease, a recessive glycolipid storage disease, found most commonly in Jewish populations. Until recently, it had been assumed that everyone who inherited two mutant copies of the gene for this condition would suffer from the disease. But, with the ability to test families where the disease is clearly inherited for the mutant gene, it appears that perhaps as many as 1/2 of people with particular mutant genes do not exhibit symptoms of the disease (Sidransky and Ginns, 1993; Beutler, 1992). Although not as striking, similar findings have been made with cystic fibrosis (Meschede et al., 1993) and Huntington's Disease (Rubinsztein et al., 1996).

This substantial variability remains unexplained. In some cases, it could be due to the effects of other genes within each individual on the expression of the phenotype; in others, it could be due to environmental factors—prenatal conditions, diet, etc. In the case of Huntington's Disease, variation in onset and severity of the disease can in some instances be explained by the surprising and novel finding that the gene responsible changes from generation to generation (The Huntington's Disease Collaborative Research Group, 1993). That is, a part of the gene amplifies in length as it is passed on from parent to child. The greater the amplification, the greater the chance of the disease being manifested, and the age at which it is contracted may also be affected.

The field of psychiatric genetics has been dramatically altered by new gene mapping technologies. For the last few years, a number of research groups have sought to locate genes correlated with manic depressive illness or schizophrenia. Several research articles claiming to have located such genes have subsequently been retracted or countered by other evidence (Baron et al., 1987; Egeland et al., 1987; Sherrington et al., 1988; Robertson, 1989; O'Donovan and Owen, 1992; Baron et al., 1993; Watt and Edwards, 1991; Owen, 1992). These failures have led some to suggest that these conditions are oligogenic (influenced by several genes) and that approaches that assume this will be more successful. Yet, we are still without a verified example of a specific gene or genes associated with these conditions.

This continuing difficulty has led to the development of ever stricter criteria for carrying out such studies and a good deal of reevaluation of approaches (Baron et al., 1990; Lander and Kruglyak, 1995; Risch and Botstein, 1996). In one of many review articles that attempt to understand the difficulties and propose directions for the future, Neil Risch and David Botstein (1996) state: "The distress engendered by the numerous reversals and non-replications has led many to rethink the paradigms being employed." They go on to suggest that "the genetic mechanism underlying the disease [MDI] in these families is more complicated than postulated...." Interestingly, the "rethinking" suggested appears to include only focusing on genetic mechanism, rather than including evaluation of possible environmental and gene-environmental interactions and covariance effects that contribute to the difficulties in locating genes.

The point here is not to argue that genes will not be found correlated with certain behaviors. Personally, I believe it very likely that at least for some cases of these psychiatric disorders and quite possibly for some examples of individual capabilities, genes will be found. Rather, what is important to learn from these experiences is that the relationship between genes and phenotype is often much more complex than scientists anticipated. It is likely that many surprises await us.

One last and perhaps most important point about the complexity of making arguments about genetics. A trait or behavior can be highly correlated with genetics, but, in fact, turn out to be largely due to environmental factors. Since the focus of this talk is on genetics and racism, I will give an example from that area. What is beyond debate is the fact that scores on tests of cognitive ability are on average correlated in this country with genes for skin color. But a recent study by Steele and Aaronson (1995) highlights the complex interaction between social, psychological, historical and other factors that mix into the achievement of scores on such a test. These researchers administered tests of cognitive ability to mixed groups of black and white college students. In some cases they told the students that the tests would measure their abilities and in others that these tests were simply problem-solving tasks that were "nondiagnostic" of ability. To quote their conclusions: "Blacks underperformed in relation to whites in the ability-diagnostic condition but not in the nondiagnostic condition." This extraordinary finding was attributed to "stereotype vulnerability."

The main reason for including this example is to highlight the extraordinary primitiveness of our real knowledge of how human behavior works and human aptitudes are manifested. There are huge unexplored realms of human behavior that we probably cannot even imagine at this point. The likely intricate interplay of so many societal and familial and genetic factors with the ultimate score that is achieved on a test by an individual will not be usefully reduced to the specification of the location of a gene on a chromosome. Those behavioral geneticists who attempt to quantify family environment in cognitive ability studies by counting the number of books in the home and assessing parental vocabulary, etc. (Rowe, 1994) are taking a remarkably simplistic view of the interactions between human development and the environment.

The lessons from genetics are then not lessons of fixity. Saying that there is a gene mutation associated with some trait rarely tells us anything about how much that trait can be changed or how much it will vary. Thus there is no basis, for example, for Herrnstein and Murray to argue from scientific evidence (on the heritability of cognitive ability) that is faulty to begin with that there is not much we can do to change the gaps in performance on test scores between groups.

## The Social Impact of Misrepresentation of Genetics

False claims for the genetic basis of human behaviors and capabilities and misrepresentation of the significance of genetics have been with us since the founding of the field. These claims have had significant social consequences as is illustrated by the eugenics movement in this country at the turn of the century and in Germany both before and during the Nazi era. More recently, the claims of Arthur Jensen (1969) for genetic evidence that blacks were inferior to whites in intelligence influenced social policy. His article from the *Harvard Educational Review* was read into the *Congressional Record* during debates over compensatory educational programs and was cited in court cases regarding school integration. Charles Murray, coauthor of *The Bell Curve*, was invited to speak to the newly elected Republican congressmen after the 1994 election.

Given past experience, arguments for the genetic basis of cognitive abilities are as likely to influence either overtly or subconsciously debates over affirmative action and educational policy. Moreover, scientific claims for the inferiority of a particular ethnic or racial group cannot help but strengthen racist attitudes. Perhaps more tragically, these arguments affect feelings of self-worth among people within those groups. For instance, in 1980, a study was published in *Science* magazine implying that girls were biologically inferior in math ability to boys (Benbow and Stanley, 1980). The publicity that that message received had a direct influence on public attitudes on the issue and affected girls' own feelings about their potential to do well in math (Beckwith, 1983). This came at a time when the percentage of women who were entering the field of mathematics was increasing dramatically.

Thus, it appears to me to be incumbent on geneticists to contribute to the public discussion about the social implications of their science for several reasons: because of the historical precedents of misrepresentation and misuse of genetics, because of the recent recurrence of these arguments, because of the clear destructive social impact of such unchecked genetic claims, and because of the influence of media coverage of contemporary genetic research including the Human Genome Project on public concepts of genetic meanings. Furthermore, since contemporary genetic research is highlighting more strongly than ever the complexity of the connection between genes and phenotype, geneticists can point out the extraordinarily simplistic and erroneous nature of the arguments put forth by academics such as Herrnstein and Murray. They can also contribute to a better representation of the meanings of genetics to the public by more carefully presenting their own results to the media and helping to correct mistaken popular representations of genetics. Speaking from personal experience, I believe that, in order for geneticists to develop a sense of their responsibility in these areas, a knowledge of the history of the public interfaces of genetics is important. Unfortunately, our education as scientists does not prepare us in any way for these problems by presenting this history.

A recent effort to respond to the genetic arguments presented in *The Bell Curve* illustrates the problems in mobilizing the scientific community to deal with these issues. The Working Group on Ethical, Legal and Social Implications of the Human Genome Project (ELSI) was established by Jim Watson to anticipate potential harmful consequences of the project and to propose policy options for dealing with them. When *The Bell Curve* appeared, the ELSI group agreed that the book represented a misuse of genetics and that geneticists should respond to it. A statement was drafted that recognized the enormous power of the genetic metaphor in society today—a power that results in part from the successes of the Human Genome Project (Nelkin and Lindee, 1995). As a group associated with that project, we in ELSI felt it necessary to point out the fallacies underlying the genetic arguments of *The Bell Curve* and their consequent applications to social policy. We prepared a statement that (after some delay) was published in the *Human Genome News* (ELSI, 1996), and subsequently in the *American Journal of Human Genetics* (Allen et al., 1996) and received attention in *Science* (Andrews and Nelkin, 1996) and *Nature* (1995) magazines.

At the same time, ELSI approached four professional genetics organizations: the American Society of Human Genetics, The American Society of Medical Genetics, The Genetics Society of America and the National Society of Genetic Counselors. The only society that agreed to endorse the statement was that composed of genetic counselors. None of the other contacts indicated that their boards necessarily disagreed with the statements, but rather responded with comments indicating that they were not familiar enough with the issues. While I would have liked it, I am not saying that these organizations should have endorsed the specific statement that ELSI prepared. But, at the least, they might have come up with a statement of their own if they were not in full accord with all of the wording. Their failure to do so is an indication of how far we have to go. More than ever, genetics is entering into the public realm. The establishment of ELSI by Jim Watson is a powerful indication of that fact. It is time that scientists become familiarized with historical analysis of the interface between science and society and begin to fulfill a function at that interface.

**References**

Anonymous. 1995. Genome research risks abuse, panel warns. Nature 378:529.

Abbott, A. 1996. Complexity limits the powers of prediction. Nature 379:390.

Allen, A., B. Anderson, L. Andrews, J. Beckwith, and et al., 1996. *The Bell Curve:* Statement by the NIH-DOE joint working group on the ethical, legal, and social implications of human genome research. Am. J. Human Gen. 59:487-488.

Allen, G. 1975. Genetics, eugenics and class struggle. Genetics 7: 2945.

Andrews, L.B., and D. Nelkin. 1996. *The Bell Curve:* A statement. Science 271: 13-14.

Bailey, M., and R. Pillard. 1991. A genetic study of male sexual orientation. Arch. Gen. Psych. 48:1089-1096.

Baron, M., N. Risch, R. Hamburger, B. Mandel, S. Kushner, M. Newman, D. Drummer, and R. H. Belmaker, R.H. 1987. Genetic linkage between X-chromosome markers and bipolar affective illness. Nature 326:289-292.

Baron, M., J. Endicott, and J. Ott. 1990. Genetic linkage in mental illness. Limitations and prospects. Brit. J. Psychiatry 157:645-655.

Baron, M., N. F. Freimer, N. Risch, B. Lerer, J. R. Alexander, R. E. Straub, S. Asokan, K. Das, A. Peterson, J. Amos, J. Endicott, J. Ott, and T. C. Gilliam. 1993. Diminished support for linkage between manic depressive illness and

X-chromosome markers in three Israeli pedigrees. Nature Genet. 3:49-55.

Beckwith, J. 1983. Gender and math performance: does biology have implications for educational policy? J. Educ. (Boston Univ.) 165:158-174.

Beckwith, J. 1987. Genetics at the Institut Pasteur: substance and style. ASM News 53:551-555.

Beckwith, J. 1993. A historical view of social responsibility in genetics. BioScience 43(5):327-333.

Bell, R., and E. Masood. 1996. 'Race and IQ' psychologist inquiry over teaching conduct. Nature 381:105.

Benbow, C., and J. Stanley. 1980. Sex differences in mathematical ability: Fact or artifact? Science 210:1262-1264.

Beutler, E. 1992. Gaucher Disease: new molecular approaches to diagnosis and treatment. Science 256:794-799.

Billings, P.R., J. Beckwith, and J. S. Alper. 1992. The genetic analysis of human behavior: a new era? Soc. Sci. Med. 35:227-238.

Brunner, H.G., M. Nelen, X. O. Breakefield, H. H. Ropers, and B. A. van Oost. 1993. Abnormal behavior associated with a point mutation in the structural gene for monoamine oxidase A. Science 262, 578-583.

Butler, D. 1995. Geneticist quits in protest at "genes and violence" claim. Nature 378:224.

Cowlet, G., and C. Hall. 1993. The genetics of bad behavior. Newsweek (Nov. 1).

Egeland, J.A., D. S. Gerhard, D. L. Pauls, J. N. Susex, K. K. Kidd, C. R. Allen, A. M. Hostetter, and D. E. Housman. 1987. Bipolar affective disorder linked to DNA markers on chromosome 11. Nature 325:783-787.

ELSI.©1093 1996. ELSI working group responds to *The Bell Curve*. Human Genome News January-March, 16.

Gould, S.J. 1981. The Mismeasure of Man. W.W. Norton & Co., New York.

Hamer, D.H., S. Hu, V. L. Magnuson, N. Hu, and A. M. L. Pattatucci. 1993. A linkage between DNA markers on the X chromosome and male sexual orientation. Science 261:321-327.

Hearnshaw, L.S. 1979. Cyril Burt, Psychologist. Hodder and Staughton, London.

Herrnstein, R .J., and C. Murray. 1994. The Bell Curve. Free Press, New York.

Jaroff, L. 1989. The gene hunt. Time March 20, 67.

Jensen, A.R. 1969. How much can we boost IQ in scholastic achievement? Harv. Ed. Rev. 33:1-123.

Kamin, L. 1974. The Science and Politics of I.Q. Potomac:Earlbaum Associates.

Kevles, D. 1985. In the Name of Eugenics: Genetics and the Uses of Human Heredity. University of California Press, Berkeley.

Kuhn, T. S. 1970. The Structure of Scientific Revolutions. 2nd Ed. The University of Chicago Press, Chicago, IL.

Lander, E., and L. Kruglyak. 1995. Genetic dissection of complex traits: guidelines for interpreting and reporting linkage results. Nature Genet. 11:241-247.

LeVay, S. 1991. A difference in hypothalamic structure between heterosexual and homosexual men. Science 25:1034-1037.

Longino, H.E. 1990. Science as Social Knowledge. Princeton University Press, Princeton, New Jersey.

Ludmerer, K. 1972. Genetics and American Society. Johns Hopkins University Press, Baltimore.

Mackintosh, N.J. 1996. Science struck dumb. Nature 381:33.

Maddox, J. 1993. Has nature overwhelmed nurture? Nature 366, 107.

Mehler, B. 1994. In genes we trust. When science bows to racism. Reform Judaism Winter:10-79.

Meschede, D., A. Eigel, J. Horst, and E. Nieschlag. 1993. Compound heterozygosity for the deltaF508 and F508C cystic fibrosis transmembrane regulator (CFTR) mutations in a patient with congenital bilateral aplase of the vas deferens. Am. J. Human Gen. 53:292-293.

Muller-Hill, B. 1988. Murderous Science: Elimination by Scientific Selection of Jews, Gypsies and Others, Germany 1933-1945. Oxford University Press, Oxford:

Nelkin, D. and M. S. Lindee. 1995. The DNA Mystique: The Gene as a Cultural Icon. W. H. Freeman, New York.

O'Donovan, M. C. and M. J. Owen. 1992. Advances and retreats in the molecular genetics of major mental illness. Ann. Med. 24:171-177.

Owen, M. J. 1992. Will schizophrenia become a graveyard for molecular geneticists? Psychol. Med. 22:289-293.

Proctor, R. 1988. Racial Hygiene: Medicine Under the Nazis. Harvard University Press, Cambridge.

Provine, W. B. 1973. Geneticists and the biology of race crossing. Science 182:790-796.

Rensberger, B. 1976. Briton's classic I.Q. data now viewed as fraudulent. The New York Times, November 28, 1.

Risch, N., and D. Botstein. 1996. A manic depressive history. Nature Genet. 12:351-353.

Robertson, M. 1989. False start on manic depression. Nature 342:222.

Rowe, D. C. 1994. The Limits of Family Influence: Genes, Experience and Behavior. The Guilford Press, New York: p.32.

Rubinsztein, D. C., J. Leggo, R. Coles and et al. 1996. Phenotypic characterization of individuals with 30-40 CAG repeats in the Huntington Disease (HD) gene reveals HD cases with 36 repeats and apparently normal elderly individuals with 36-39 repeats. Am. J. Human Gen. 59:16-22.

Sherrington, R., J. Brynjolfsson, H. Petursson, M. Potter, K. Dudleston, B. Barraclough, J. Wasmuth, M. Dobbs, and H. Gurling, H. 1988. Localization of a susceptibility locus for schizophrenia on chromosome 5. Nature 336:

164-167.

Sidransky, E., and E. I. Ginns. 1993. Clinical heterogeneity among patients with Gaucher's Disease. J. Am. Med. Assoc. 269:1154-1157.

Steele, C.M., and J. Aronson. 1995. Stereotype threat and the intellectual test performance of African Americans. J. Personality Soc. Psychology 69:797-811.

Strohman, R.C. 1993. Ancient genomes, wise bodies, unhealthy people: limits of a genetic paradigm in biology and medicine. Perspectives Biol. Med. 37: 112-145.

The Huntington's Disease Collaborative Research Group. 1993. A novel gene containing a trinucleotide repeat that is expanded and unstable on Huntington's Disease chromosome. Cell 72:971-983.

Tourian, A. and J. B. Sidbury. 1983. Phenylketonuria and hyperphenylalaninemia. In: The Metabolic Basis of Inherited Disease, edited by Stanbury, J.B., Wyngaarden, J.B., Frederickson, D.S., Goldstein, J.L. and Brown, M.S. New York: McGraw-Hill, pp. 270-286.

Watt, D.C., and J. H. Edwards. 1991. Doubt about evidence for a schizophrenia gene on chromosome 5. Psychol. Med. 21:279-285.

Wolf, U. 1995. The genetic contribution to the phenotype. Hum. Genet. 95:127-148.

# ASSESSING THE HUMAN GENOME PROJECT:
## An African-American and Bioanthropological Critique

### Fatimah Jackson

Forty years ago, Macon County, Alabama, was the site of a well-orchestrated set of scientifically interesting but morally bankrupt investigations by the U.S. Public Health Service to purportedly ascertain the natural history of syphilis. I am speaking of course about the Tuskegee Syphilis Study, Tuskegee being the Macon County seat. The fact that African-American men and their families in this county were the primary victims of this scientific study was no accident. American science has a long history of using its most politically and socially disenfranchised citizens in experimental biomedical endeavors. Once the knowledge gained from field and laboratory observations is quantified and analyzed, and the resulting innovations standardized, we usually see a shift in the target of group interest from low status African-Americans to more well enfranchised groups. Once the beneficial aspects are clarified and their harmful side effects relatively minimized, only then do we find that access to the products of scientific insight are made available to the higher castes in American society and largely denied to the "less privileged castes." In light of this historical pattern in applied biomedical science in this country, how should African-Americans and other American citizens now view the largest and most costly taxpayer-funded scientific effort since the Apollo space mission, the Human Genome Project?

Advances in DNA technology have made widespread genetic testing highly feasible and virtually inevitable. DNA sequencing of the human genome under the aegis of the Human Genome Project is expected to be completed before 2000 CE. This is largely the result of the application of automated DNA sequencing, improved sampling handling, advanced bioinformatics, and other technological innovations to the project. Additionally, an increased number of researchers are devoting their most productive career years to genomic research, further accelerating the rate of identification and description of important genomic sequences.

The author is a professor and Distinguished Scholar Teacher in the Department of Anthropology at the University of Maryland, College Park, MD 20740 where she heads the Genomic Models Research Group.

For many accustomed to looking for the "big picture," the Human Genome Project is predicted to bolster two truisms of modern human variation: (1) that there exists a high degree of homogeneity (genetic similarity) between members of various human geographical groups (for example, Africans, Asians, and Europeans) and (2) that there are also high levels of heterogeneity within such large geographical groups and their re-sorted descendants in the Western Hemisphere. Many recognize that the molecular genetic data hold the promise to significantly challenge the existing folk paradigms concerning the origin, meaning, and importance of human biodiversity. However, it is also recognized that these same data, in the wrong hands and with the wrong motivations, can concretize current socioeconomically-rooted inequities under a thin veneer of "science," as the history of American eugenics makes abundantly evident.

Given this backdrop, it is not clear how the Human Genome Project, as currently conceived and implemented, will directly benefit African-Americans. Primarily this is because an important, virtually undiscussed scientific contradiction exists between the methods used for sampling and the stated goals of the project. It is this important contradiction that provides a foundation for this critique of the HGP and a basis for the subsequent suggestions for improving the project's scientific accuracy and relevance to all of humanity.

The Human Genome Project is the most important molecular taxonomic effort of this century. The HGP will ostensibly define, at the molecular genetic level, the human being. Since all modern humanity is a single species—indeed a single subspecies, by setting the taxonomic norm for *Homo sapiens sapiens,* the HGP will establish a baseline from which all future molecular comparisons (e.g., data from the Human Genome Diversity Project or HGDP) will be made. But many have asked, *"Who* will comprise this baseline?" The overwhelming majority of the genes being sequenced for the Human Genome Project come from a very limited number of North Atlantic European-American lineages. No serious attention has been paid to the issue of the representativeness of these lineages to the rest of humanity and their appropriateness for forming the taxonomic baseline for the human genome.

If, as Eric Lander suggests (1996), the HGP will provide the "structural periodic table" from which functional genomics will emerge, we must ask, "will the variation most frequently encountered among African-Americans be reflected in this molecular periodic table?" Will African-Americans and other non-Europeans serve as something more substantive than a colorful and diverse backdrop to the current narrow slice of North Atlantic European diversity that is the focus of the Human Genome Project? Right now, the bulk of the HGP genes have come from a few European-American families whose cell lines were opportunistically available for distribution and genomic analysis. How will the HGP accurately reflect the true temporal (i.e., historic) or spatial (i.e., geographic) patterns of biodiversity that characterize modern humanity?

Since the Human Genome Project is *refining* (and redefining) our notions of human microtaxonomy by establishing a molecular-based norm, what criteria are HGP scientists using in identifying which alleles should contribute to this taxonomic norm? How will Human Genome Project information benefit Americans of African descent and other citizens of color who have shared in the expense and sacrifice for this highly-touted project? How will their inevitable variation from the eurocentric molecular baseline generated by the HGP be addressed? Will variation from the molecular periodic table be further grounds for stigmatization? How will molecular variants frequently observed in African-Americans be categorized? In some cases, these very variants may more closely represent original genes rather than their derivations. Will the HGP and the biomedical scientists hoping to improve their diagnostic capabilities with HGP-generated data be able to distinguish "wild types" from mutants? Will future taxonomic assessments based upon evolutionarily peripheral lineages be skewed?

The Human Genome Project aims to reflect the normal sequence for 100,000 genes and accurately depict the ancestral and functional affinities among these genes. However, such accuracy is only feasible with more precise bioanthropological and biohistorical considerations going into the selection of the groups and individuals whose genes will comprise this genomic "periodic table." Although few anthropological geneticists, biological anthropologists, and human population biologists have been included in the HGP "inner circle," it is clear from even a cursory review of these disciplines that there are systematic ways of ensuring fairness and representativeness in the Human Genome Project sample and hence greater applicability of the HGP data to a larger contingent of humanity. General guiding principles from population biology suggest several appropriate alternative sampling strategies:

(1) Identify sequences from the oldest (i.e., deepest) lineages of our species to obtain the most commonly shared variants among all of humanity. This sampling strategy would mandate the selection of alleles from lineages in eastern and southern Africa where the oldest anatomical, molecular, and behavioral evidence for modern humans exists.

(2) Collect sequences from the most frequently encountered contemporary phenotype(s) to increase the likelihood that any derived sequences would at least be common among contemporary humans. This sampling strategy would encourage researchers to focus on the most commonly encountered contemporary human phenotypes, ethnic Han Chinese and East Indian females.

(3) Sample proportionally from each geographical region of the human range based upon current population densities to ensure that the final genome is inclusive. Specifically, this sampling strategy is for proportional random sampling based upon current population distributions; and, finally,

(4) Sample probabilistically on an international basis. Although randomized, this sampling strategy would have the potential to oversample sparsely populated areas of the human range and undersample more dense regions or more biodiverse groups. The fact that none of these basic strat-

egies are being employed in the selection of sequences for HGP analysis will severely limit the broad scientific and biomedical utility of this massive technological effort as well as weaken its taxonomic accuracy (Jackson, 1997a, 1997b). And, for African-Americans, the products of the current HGP population sampling may, in fact, do harm.

Human Genome Project backers dazzled congressional funders with the vision of identifying (and manipulating) disease genes of broad public health significance. Yet, our current repertoire of disease genes and genetic polymorphisms in general is highly biased toward European patterns of diversity, especially North Atlantic European groups (Cavalli-Sforza et al., 1994). No studies suggest that these polymorphisms are representative for all of humanity. The selected genes identified by the Human Genome Project to comprise the human "norm" may not necessarily be the most relevant genes for non-European or even other non-western European groups. This fact significantly limits our current genetic "toolkit."

Aside from the recurrent socioeconomic issues of stigma and disenfranchisement associated with variation from the "norm," recent advances in the genetics of human disease have caused many researchers to modify our old working models of genetic diversity. In a recent summary of these advances, Weiss (1996) notes that new DNA sequence data have shown that most mutations are unique (rather than recurrent), that there are many (tens, even hundreds, not just two) alleles at a typical locus, and that variation in any geographic area is composed of clones of related alleles (therefore, cases of a given disease in a specific geographic area are usually genetically related; this same disease in another geographic area may involve a different set of genetically related alleles).

For example, the specific alleles for cystic fibrosis that are common in Europeans are not necessarily the same alleles responsible for cystic fibrosis in non-European groups (Tsui, 1992; CFGAC, 1994). Therefore, screening and gene therapy intervention strategies that are appropriate for one geographic group may prove ineffective for another group afflicted with the same phenotypic disease condition. A genetic map based on one group may be essentially useless for identifying and depicting the genes relevant for the same disease state in another geographically distinct group.

Even within geographic groups, significant variation occurs. Extensive molecular genetic studies of the Finns (Proc. Natl. Acad. Sci., 1996) suggests that they are highly homogeneous, relative to other Europeans, and exhibit an exceptionally low frequency of cystic fibrosis relative to neighboring European groups. In this context then, a Finn-based map of the human genome would not even provide much insight into European disease-gene relationships.

Clearly, population history has had a major impact on the patterns of human genetic variation of biomedical importance. What will be the real potential for identifying the most relevant disease-related genes for use in evaluating disease causation across geographical groups? If identification is based on a taxonomically non-representative set of alleles, will we have in the Human Genome Project the most appropriate disease-gene sequences to facilitate the development of highly specific diagnostic and therapeutic products for the majority of humanity? The use of nonrepresentative markers may not accurately reflect the precise patterns of relationship among and between contemporary geographical groups or provide the most relevant and safe therapeutic interventions. Therefore, our search for relevant disease genes in the context of the HGP must be extended to include other world groups.

Currently, the HGP seems to suffer from an unquestioned, unexamined, spontaneous overextrapolation of European-American (and male) patterns as human universals. This is not new in American science. Indeed, although many social scientists have begun to investigate the social construction of "Whiteness as universal," much biomedical research continues to project western Europe and European-American standards as the monothematic world norm. This minimizes diversity within groups and fosters the abundance of simple (and erroneous) biomedical and epidemiological comparisons between members of highly diverse groups. In America, this literature is replete with comparisons of "Blacks" and "Whites" that are not only unspecific from a genetic perspective but are also of limited diagnostic utility and of even less comparative value. Invariably in such comparisons, "Blacks" are designated pathological or at best seen as deviant relative to the implicit "White" norm. If the Human Genome Project does not extend itself to other world groups in a scientifically and morally justifiable way, it too will likely exacerbate, rather than reduce, the "Black-White" dichotomization in American biomedicine and epidemiology and the continued, scientifically sanctioned, disempowerment of African-Americans.

African-Americans have long recognized (and appreciated) the high degree of social, cultural, and biological variability within this socially-constructed macroethnic group. Much of this variability reflects regional diversity in west and central African origins, coupled with modest variable degrees of admixture with specific local groups of non-African-Americans. Table 1 summarizes the historical foundations for contemporary African-American biodiversity.

Indeed, throughout the United States and elsewhere, large, highly variable, recently admixed groups such as "Blacks," "Whites," "Hispanics," "Asians," and "Native Americans" often show regional patterns of diversity. These regional patterns reflect variation in the intensity of specific microevolutionary forces (e.g., natural selection, genetic drift, gene flow, nonrandom mating). A significant cause of regional diversity within such groups has been the different ancestral groups contributing regional group variation in the rate of gene flow between such groups. Given the resulting heterogeneity of such macroethnic groups, how can the small group of North Atlantic European-American lineages that currently dominate the HGP genomic sample be expected to adequately reflect the wealth of diversity within such macroethnic groups?

Finally, when and if African-Americans are systematically included into "biology's periodic table"—the Human Genome Project, what will be the associ-

Table 1. Summary of African regional origins of various local African-American groups and summary of major non-African groups historically admixing with African-Americans at specific important sites.

| US Importation Site | Location | African Regional Origins | Percentage/Site |
|---|---|---|---|
| Chesapeake Bay | ATLANTIC | Bight of Biafra | 38% |
| | OCEAN | Central Africa | 16% |
| | INLETS IN | Gold Coast | 16% |
| | VIRGINIA | Senegambia | 15% |
| | AND | Upper Guinea | 11% |
| | MARYLAND | Mozambique | 4% |
| | | | 100% |

| Sources for Admixture with African-Americans | |
|---|---|
| Major nearby Native American groups: | Major colonizing European groups: |
| Delaware, Piscataway, Nahyssan, | Ulster English, Irish, Scots, |
| Nanticoke, Powhatan, Monacan, | German, Scots-Irish, Welsh |
| Pamunkey, Meherrin, Choptank, | |
| Saponi, Tutelo, Susquehanna | |

| US Importation Site | Location | African Regional Origins | Percentage/Site |
|---|---|---|---|
| Carolina Coast | BAY AND | Central Africa | 40% |
| | LOWLAND | Senegambia | 23% |
| | AREAS OF | Upper Guinea | 18% |
| | NORTH AND | Gold Coast | 9% |
| | SOUTH | Bight of Biafra | 7% |
| | CAROLINA | Bight of Benin | 3% |
| | | | 100% |

| Sources for Admixture with African-Americans: | |
|---|---|
| Major nearby Native American groups: | Major colonizing European groups: |
| Nansemond, Coratan (Lumbee), Creek, | French Huguenots, Highland Scots, |
| Cherokee,Tuscarora, Nottaway, | English Quakers, Scots-Irish, Swiss, |
| Tutelo, Neusiok, Sewee, Catawba, | Palatinate Germans, Spanish |
| Waccamaw, Pamlico, Hitchiti, Muskogee | |

| US Importation Site | Location | African Regional Origins | Percentage/Site |
|---|---|---|---|
| Mississippi Delta | GULF AREAS | Senegambia | 32% |
| | OF TEXAS, | Central Africa | 25% |
| | LOUISIANA, | Bight of Benin | 25% |
| | MISSISSIPPI | Bight of Biafra | 8% |
| | AND INLAND | Upper Guinea | 6% |
| | | Gold Coast | 2% |
| | | Mozambique | 2% |
| | | | 100% |

| Sources for Admixture with African-Americans | |
|---|---|
| Major nearby Native American groups: | Major colonizing European groups: |
| Chitimacha, Biloxi, Choctaw, Ofo, | Acadia French, Spanish |
| Houmas, Tunica, Chickasaw, Caddo, | |
| Coushatta, Atakapa, Karankawa | |

Data derived from: Jackson et al., 1996a,b

ated benefits and liabilities of being part of the data base? WIll inclusion be coupled with improvements in the provision of health and education services to the African-American community or will inclusion simply facilitate, at a molecular level, the coercive and oppressive aspects of previous "bad science" on this group?

In 1994, a consortium of biological and social scientists met in Washington, D.C. to draft the *African American Manifesto on Genomic Studies.* The group was convened in part to rectify the misrepresentation of African-American perspectives and priorities articulated by a representative of the Human Genome Diversity Project (HGDP) at the World Council of Indigenous Peoples Conference in Quezaltenango, Guatemala in December of the previous year. Given the topic of this paper, it is perhaps relevant to reiterate the most salient conclusions of that document here. The six key points of the *Manifesto* were:

(1) African-Americans expect full inclusion in any world survey of human genomic diversity. African-Americans represent an amalgamation of African peoples with a unique lineage and cultural history. The inclusion of African-Americans is not optional in any world survey, particularly if U.S. taxpayer monies are the funding source for such efforts.

(2) Given the high degree of genetic heterogeneity already uncovered among African-Americans, no scattered, opportunistic samples can be validly extrapolated to the entire group of African-Americans. It is imperative that systematic sampling, either model-based or design-based (probability), be used to identify the broad range of variability existing among African-Americans and that this diversity be linked to other relevant social, cultural, historical, and ecological features of the African-American existence.

(3) We are working toward the establishment of a National Review Panel for ongoing evaluations of genomic studies among African-Americans. This National Review Panel will issue certification of projects that are consistent with the research aims and objectives of the African-American community. The National Review Panel will also identify genetic research projects that are not consistent with the research needs of the African-American community and may, in fact, be harmful to the community.

(4) As part of the certification process, acceptable genomic studies among African-Americans must include representatives of the African-American community in all aspects of the research process. African-Americans must participate in the research design, research implementation, data collection, data analysis, data interpretation and dissemination of research results (i.e., scientific publications and policy implications of the results).

(5) A priority will be given to genomic studies that examine the linkage of African-Americans to continental Africans and other Africans of the various diasporas. For too long, research dollars have been squandered on meaningless "black-white" comparisons, most of which provided few positive policy results for the African-American community. The historical circumstances of African-Americans has made it important to many in our community to identify the

genetic connections between African-Americans and Africans elsewhere. Therefore, this community priority must significantly inform prospective genomic research efforts.

(6) Genomic sampling of African-Americans must be linked to improvements in the provision of health and educational services to the African-American community. The largest proportion of health problems in the African-American community are due to disorders that are preventable. These preventable environmental and gene-environment diseases should be addressed in conjunction with genomic studies. Indeed, the target of both disease prevention and genomic studies among African-Americans must be the improved health and well-being of the population and its enhanced survival into the coming centuries. Groups and individuals wishing to conduct genomic sampling among African-Americans are expected to concurrently provide meaningful educational and training opportunities for African-Americans.

The Human Genome Project's scientific unrepresentativeness is its major limitation from both African-American and bioanthropological perspectives. The issue of unrepresentativeness is neither trivial nor academic. Once the molecular taxonomic norm is in place, it becomes very difficult to dislodge. Additionally, it provides the template for all subsequent biomedical and evolutionary assessments of human biodiversity.

On the other hand, the most important scientific value of accurate, population-representative molecular genetic studies lies in the potential for generating data that gives insight into the true biohistoric and phylogenetic relationships of contemporary and ancient human groups and the development of broadly efficacious gene therapies. Well-designed studies based upon a correct taxonomic norm can yield a better understanding of human origins and distinguish between original (identified as "primitive") traits versus subsequent mutations (identified as "derived" traits). Anthropologically-informed molecular studies can identify the degree of within-group diversity and yield a better understanding of gene-gene and gene-environment interactions. These interactions can lead to sound gene-based preventive and therapeutic regimes that are specific to various groups of humans. Understanding the interactive foundations of disease expression is extremely important since most gene-influenced diseases (e.g., diabetes, hypertension, various cancers) represent complex phenotypes and are rarely the result of single gene expression.

Poorly designed molecular genetic studies cannot be broadly extrapolated, regardless of the rigor of the bench chemistry. DNA research that focuses only on a "universal norm" that is at variance with the true patterns of human biohistory and/or contemporary biodemography is based on a lie and ultimately imprisons all who stand outside that narrowly configured (and often self-serving) "norm." In DNA studies minimizing (or comparing) variation within a subspecies (even within a species), such as that being undertaken by the HGP, careful science must precede the selection of populations and individuals for study. This is essential because of the high degree of genetic similarity between groups. Any variation between geographic groups will be, at best, at the microevolutionary level (that is, occurring below the species level). If the Human Genome Project is to be directly relevant to African-Americans, it must shift to a new, inclusive structure. Otherwise, the functional genomics generated will be of limited utility for humanity, the Human Genome Project misnamed, and African Americans and others will have been asked once again to support a project that benefits a privileged few.

**References**

Cavalli-Sforza, L. L., P. Menozzi, P. and A. Piazza. 1994. The History and Geography of Human Genes. Princeton Univ. Press: Princeton, NJ.

CFGAC (Cystic Fibrosis Genetic Analysis Consortium). 1994. Population variation of common cystic fibrosis mutations. Human Mutation 4:167-177.

Jackson, F. L. C. 1998. Scientific limitations and ethical ramifications of a non-representative Human Genome Project: African American responses, Science and Engineering Ethics. (In press)

Jackson, F. L. C. 1997a. The taxonomic implications of the HGP. American Journal of Human Biology 9(1):130.

Jackson, F. L. C. 1997b. Concerns and priorities in genetic studies: Insights from recent African American biohistory. Seton Hall Law Review 27(3):951-970.

Jackson, F. L. C., K. R. Reed, T. L. Tetrault, and M. G. Franke. 1996a. Ethnogenetic layering of the Crescent States: using a model-based strategy for population genetic sampling. American Journal of Human Biology 8(1):118.

Jackson, F., K. Reed, G. Franke, and T. Tetrault. 1996b. Regional diversity in African origins of African Americans within the Crescent States. Program Abstracts, Tuskegee University Conference on the Human Genome Project, September 26-28, Tuskegee, AL.

Lander, E.A. 1996. The new genomics: global views on biology. Science (Oct. 25) 274:536-539.

Tsui, L-C. 1992. The spectrum of cystic fibrosis mutations. Trends in Genetics 8:391-398.

Weiss, K.M. 1996. Is there a paradigm shift in genetics? Lessons from the study of human diseases. Molecular Phylogenetics and Evolution 5(1):259-265.

# G-RAP: A MODEL HBCU GENOMIC RESEARCH AND TRAINING PROGRAM

**Georgia M. Dunston**

In this closing decade of the 20th century, the Human Genome Project has extended the probing of biomedical science to the ultimate level of biological identity, i.e., unique DNA sequence variation. Moreover, exploration of the human genome has introduced new prospects for understanding the molecular processes underlying disease, disease susceptibility, and genome evolution. The advent of recombinant DNA technology has had a profound effect on the growth of human genome research and the progress of the project.

The phenomenal success of the Human Genome Project in the construction of high resolution comprehensive linkage maps for locating disease genes has precipitated an avalanche of new data on the amount and distribution of genetic variation at the genome level. The latter challenges prevailing concepts on the nature and frequency of mutations, allelic variation, haplotype polymorphisms, and the presence of genetic variations in different population groups and associated geographic distribution of disease.

This paper describes the origins of G-RAP (Genomic Research in African-American Pedigrees) and its development as *A Model HBCU (Historically Black Colleges and Universities) Genomic Research and Teaching Program*. G-RAP was proposed in 1990 contemporaneously with the beginning of the first five years (FY 1991-1995) of the U.S. Human Genome Project.

The scientific goal of the HGP, to map and sequence the human genome, has been greatly facilitated by the discovery of polymorphic microsatellite markers widely distributed throughout the genome (Weber and May, 1989; Jeffreys et al., 1990) and the availability of a common reference pedigree collection and shared genotype repository. However, use of the CEPH reference panel, a collection of 67 Caucasian families, for scientific collaboration in the construction of comprehensive high resolution genetic maps (Dausset et al., 1990) raised questions in the minds of some about the broad applicability of genomic polymorphic marker information based on this limited spectrum of the human genome.

The author is Professor and Interim Chair of the Department of Microbiology and Director of the G-RAP Program at the Howard University College of Medicine, Washington, D.C. 20059. © May not be reproduced in any form without the expressed written consent of the author.

As contemporary descendants of the oldest human populations, African-Americans, like other African populations, often show greater genomic diversity than Caucasian populations (Cavalli-Sforza et al., 1994). Since the greatest genetic distances in human population groups have been found between African and non-African people (Wainscoat et al., 1986), Howard University scientists made a case for the need to broaden the base of reference resources for the Human Genome Project by the addition of reference cells and/or DNA from African-Americans. The latter provided the motivation for the G-RAP Project.

The long-range goal of G-RAP is to improve the health status of African-Americans through research on DNA variability and the application of knowledge gained from research to better understand the biomedical significance of gene-based differences already known to exist among populations in immune response to organ transplants, susceptibility to diseases such as diabetes, sensitivity to drugs, cancer, and the influence of environment on health. The scientific objectives of G-RAP are to identify and characterize DNA polymorphic markers useful in mapping genes underlying disease or susceptibility to diseases common in African-Americans.

These objectives are related to two broad hypotheses—population variability in polymorphisms of DNA markers used to construct the map of the human genome, and correlation of this genomic variability with disease. The linkage relationships of polymorphic DNA loci, which are inherited as simple Mendelian codominant alleles, can be determined by their co-segregation in human pedigrees (Ott, 1985). The complete genetic linkage map of DNA marker loci is a valuable resource for mapping genes segregating with disease traits in human pedigrees. The distance between African and non-African population groups raised questions about the biomedical significance of differences between African-Americans and Caucasians in DNA marker polymorphisms associated with disease susceptibilities.

By addressing population variability in the human genome, G-RAP interfaces HGP interests in mapping and sequencing the genome with scientific interests in genetic variations, which is the focus of the Human Genome Diversity Project (HGDP). The latter is a collaborative research project that is being developed on a global basis under the auspices of the Human Genome Organization. The overall goal of the HGDP is to investigate the variation occurring in the human genome by studying samples collected from populations that are representative of all of the world's peoples and, ultimately, to create a resource for the benefit of all humanity and for the scientific community worldwide.

The HGDP is designed to arrive at a much more precise definition of the origins of different world populations by integrating genetic knowledge, derived by applying the new techniques for studying genes, with knowledge of history, anthropology and language. Various studies of human populations, using a variety of classical protein polymorphisms, mitochondrial DNA (Cann et al., 1987), Y chromosome DNA (Hammer, 1995), and nuclear DNA (Armour et al., 1996) are consistent with the hypothesis that the earliest divergence in human evolution separated African and non-African lineages. Understanding population genomic characteristics is necessary for optimal utilization of these highly informative loci for gene mapping studies. With emphasis on the biomedical significance of DNA sequence variability in African-Americans, G-RAP expands the context for assessing genome variability to the level of population biology and, in so doing, overlaps the scientific interests of both the U.S. Human Genome Project and the international Human Genome Diversity Project.

While significant progress is being made in the characterization of genome variations that African-Americans share in common with Caucasians, comparable progress is not being made in the characterization of genomic variation rooted in the African ancestry of African-Americans. To be properly understood and incorporated into the emerging global picture of humankind, unique genome variability in African-Americans must be identified, defined, and characterized. The latter is the focus of the Human Immunogenetics Laboratory at Howard University, which was established to investigate the biomedical significance of unique and/or common human leukocyte antigens (HLA)/gene polymorphisms in African-Americans and facilitate biomedical research of genome-wide variability in this population. The HLA system is characterized by an extensive amount and pattern of loci, allele, and haplotype polymorphisms (Bodmer et al., 1995). Since HLA polymorphisms have been associated with susceptibility to various diseases (Tewari and Terasaki, 1985), genetic variations in this system may be instructive in understanding the biology of multiple mutations, multiple loci, and microsatellite haplotype polymorphisms associated with complex disease traits.

Howard University is singular among the Historically Black Colleges and Universities in having a health sciences complex with an established Human Immunogenetics Core Laboratory for HLA and other collaborative genomic and biomedical research. In addition, Howard University has a Genetic Counseling program as one of several specializations in the graduate program in Genetics and Human Genetics. The latter includes a strong emphasis on ethical aspects of human genetics research and how human values integrate with science. Ethical, legal, and social issues are major areas of interest and concern for the HGP, HGDP and the public, especially those who are subjects in human genome testing and screening programs.

Howard University is uniquely poised, strategically positioned, and scientifically prepared to forge a national and international research agenda for G-RAP. Thus, the specific aims of G-RAP are to: (1) facilitate advancement of a national and international agenda for G-RAP; (2) investigate biomedical and biological significance of HLA and non-HLA genomic polymorphisms in African-Americans, and (3) analyze the biomedical significance of genome variability in complex diseases that disproportionately affect African-Americans.

Let me present a chronology of the development of the G-RAP program at Howard. A planning grant to develop a research program for the G-RAP project was funded in 1991 by the National Center for Human Genome Research

(NCHGR) as a Supplemental Award to the Howard University Research Centers in Minority Institutions (RCMI) Interdisciplinary Program. The planning grant supported a series of workshops addressing the four core areas of G-RAP, namely: the collection of a reference family panel representative of genome variability in African-Americans for collaborative human genome studies; establishment of a core reference resources facility for central receiving and processing of biological specimens, maintenance, quality control, and storage of reference resources; scientific utilization of reference resources; and data management (informatics).

The G-RAP planning project led to the conceptualization of a multidisciplinary, interinstitutional proposal for a "Howard University Center for Collaborative Genomic and Biomedical Research." Since 1994, efforts have been directed toward support for a G-RAP extension to the Human Immunogenetics Core Laboratory. The latter serves as a core resource for several intrainstitutional and interinstitutional collaborative studies on HLA and non-HLA genome variability in disease association and/or gene mapping studies in African-Americans. In 1995, investigators at Howard University and the NCHGR, Division for Intramural Research, began collaborative projects on mutational analyses of *BRCA1* in high-risk African-American breast cancer families, West African origins of non-insulin dependent diabetes mellitus in African-Americans and, more recently, the development of plans for a collaborative project on the genetics of prostate cancer in African-American men. These three projects are prototypes of interinstitutional collaborations that are foundational to plans underway for a human genome research center at Howard University.

Second generation HGP centers must consider the impact of population history, differences in social structure, as well as DNA sequence variability in understanding the biology of disease and disease susceptibility. Moreover, consideration must be given to ensuring informed consent and privacy of genetic information for populations tested and/or screened, especially for groups or populations that are justifiably suspicious of scientific programs. Second generation HGP centers must build upon and interpret the tremendous amount of DNA sequence variation uncovered by first generation HGP centers while en route to construction of high density genetic maps and sequencing the human genome.

The avalanche of molecular sequence data is not only impacting the rate at which genes are being mapped, isolated, and characterized, but also the way in which genomic variation is related to disease expression. An example of this impact can be seen in the recognition that multiple mutations (alleles), in contrast to conventional models of two alleles (e.g., mutant/disease vs. wild type), occur in disease genes and these alleles have a hierarchical, history-dependent structure and geographic distribution in human populations (Weiss, 1996). Thus, population history is a factor in the distribution of human genetic variation. The latter is relevant to biomedical research because a given allele is often found only in one major population, such as the specific alleles for cystic fibrosis that are common in Europeans. Other populations have their own different alleles for the "same" disease. Thus, population differences in polymorphic marker haplotypes must be considered in genetic screening and intervention.

In the organizational structure of the proposed Howard University Center for Collaborative Genomic and Biomedical Research, G-RAP is a core facility interfacing the resources of the Human Immunogenetics Laboratory with the G-RAP consortium, database, and communications network. The details of interactions between the component parts of this structure are being planned. The results of G-RAP collaborations between intramural and extramural investigators are expected to increase the depth and breadth of our collective understanding of human genome variability.

Recognizing that genome variability is the norm, rather than the "aberrant" or "mutant," recasts the way in which biology can be structured. For example, inherent diversity in the human genome may serve as a basis for defining genome-based biological relationships that can be compared at the phenotypic level relative to mechanisms of disease, disease susceptibility, and disease heterogeneity. It is perhaps ironic that reductionist thinking, which has characterized the tremendous power of modern science to successfully partition, separate, and resolve human identity to the level of DNA sequence variability, now reveals new systems of DNA polymorphisms that may provide an unbiased biological foundation for resynthesis, reconnection, and better definition of humankind. These new systems identify humanity as one continuously evolving, interdependent, and interconnected family defined by patterns of diversity inherent in all of its component parts.

Biomedical scientists at Howard University, in collaboration with other investigators in the wider scientific community, seek through G-RAP to expand both the content (i.e., resources) and context (i.e., human evolutionary biology) of our knowledge and collective understanding of DNA variability in relationship to population differences in disease susceptibilities. An understanding of genomic polymorphisms in the context of population differences is paramount in the rational design of DNA based strategies for diagnosis and treatment of human disease, as well as in the design of new and innovative genome-based technologies for health promotion and disease prevention.

## References

Armour, J. A., T. Anttinen, C. A. May, E. E. Vega, A. Sajantila, J. R. Kidd, K. K. Kidd, J. Bertranpetit, S. Paabo, and A. J. Jeffreys. 1996. Minisatellite diversity supports a recent African origin for modern humans. Nature Genet. 13(2):154-160.

Bodmer, J. G., S. G. Marsh, E. D. Albert, W. R. Bodmer, R. E. Bontrop, D. Charron, B. Dupont, H. A. Erlich, B. Mach, and W. R. Mayr. 1995. Nomenclature for factors of the HLA system. Tissue Antigens 46(1):1-18.

Cann, R. L., M. Stoneking and A. C. Wilson. 1987. Mitochondrial DNA and human evolution. Nature 325:31-36.

Cavalli-Sforza, L. L., P. Menozzi and A. Piazza. 1994. The History and Geography of Human Genes. Princeton Univ. Press, Princeton, NJ.

Cystic Fibrosis Genetic Analysis Consortium. 1994. Population variation of common cystic fibrosis mutations. Hum. Mutat. 4:167-177.

Dausset, J., H. Cann, D. Cohen, M. Lathrop, J. M. Lalouel, and R. White. 1990. Centre d'etude du polymorphisme humain (CEPH): collaborative genetic mapping of the human genome. Genomics 6:575-577.

Hammer, M. F. 1995. A recent common ancestry for human Y chromosomes. Nature 378: 376-378.

Howard University Research Centers in Minority Institutions Interdisciplinary Program. Supplement: Genomic Research in African-American Pedigrees: A Planning Project. 1991. (PI:Epps, CH/Dunston, GM), Public Health Service Grant No. G12- RR03046-07Sl.

HUGO. 1995. The Human Genome Diversity Project. Summary Document. HUGO Europe, London, UK.

Jeffreys, A. J., R. Neumann, and V. Wilson. 1990. Repeat unit sequence variation in minisatellites: a novel source of DNA polymorphism for studying variation and mutation by single molecule analysis. Cell 60:473.

National Institutes of Health. 1990. Understanding Our Genetic Inheritance . The U.S. Human Genome Project: The First Five Years FY 1991-1995. Publication No. 90-1590.

Ott, J. 1985. Analysis of Human Genetic Linkage. Johns Hopkins University Press, Baltimore and London.

Tewari, J. L., and P. I. Terasak. (Eds.). 1985. HLA and Disease Associations, Springer-Verlag, NY.

Wainscoat, J. S., A. V. S. Hill, A. L. Boyce, J. Flint, M. Hernandez, S. L. Thein, J. M. Old, J. R. Lynch, A. G. Falusi, D. J. Weatherall, and J. B. Clegg. 1986. Nature 319:491-493.

Weber, J., and P. E. May. 1989. Abundant class of human DNA polymorphisms which can be typed using the polymerase chain reaction. Am. J. Hum. Genet. 44:388-396.

Weiss, K. M. 1996. Is there a paradigm shift in genetics? Lessons from the study of human diseases. Mol. Phylogen. Evol. 5(1):259-265.

# 3

# The Perils of the Human Genome Project

# ETHICAL, LEGAL AND SOCIAL CONCERNS ABOUT THE HGP AND GENOME DATA USE: AN INTRODUCTION

## Hans S. Goerl

First of all, I want to thank Tuskegee University for organizing this conference, putting it together in very short order and arranging for some of the truly top-drawer speakers in this area. The biggest problems in genetics are those of communication and education. Conferences like these are absolutely vital in the mission of the Human Genome Project and in particular of the ELSI or Ethical, Legal and Social Implications program.

Nineteen years ago, Congressman Peter Rodino (D-NJ) was sitting in a congressional hearing and he was being asked to approve a request from NIH for a genetic program. He said at that time—bear in mind this is 19 years ago: "You know, this technology will more intimately affect our daily lives and the politics of society than any technology I know. We had better make sure early on that we have the appropriate social mechanisms in place to ensure that this technology is not abused and misused by institutions in society."

Well, almost 20 years have gone by and in those 20 years genetic scientists have succeeded in deciphering the human code to a degree far greater than they could have ever expected back then. And the technology they have created is in use today in clinical settings and in universities and genetic centers around the country. Unfortunately, almost none of this technology has been independently appraised, reviewed, or approved by the Food and Drug Administration, which historically has jurisdiction over such matters. Under pressure from Congress not to stand in the way of economic progress, that agency has abdicated from its role of independent oversight of new medical technology.

Yet, every day, people are being solicited to take predictive genetic tests that presume to tell them whether they are likely to become victims of breast cancer, Alzheimer's disease, stroke, melanoma, Huntington disease and a number of other late-onset diseases. Moreover, every day hundreds of expectant

A practicing attorney, the author is the founder and executive director of The Genethics Center (21 Summit Avenue, Hagerstown, MD 21740), an advocacy group dedicated to education about and opposition to genetic discrimination. ©May not be reproduced in any form without the expressed written consent of the author.

mothers are having genetic tests done which tell them the likelihood of some birth defects. Others are going through very sophisticated in vitro fertilization processes whereby multiple embryos are created. Those which are afflicted by genetic disease—Tay-Sachs comes to mind in particular—are discarded, and those which are healthy are implanted. Yet none of the social mechanisms of which Congressman Rodino spoke are in place today.

You will hear from our panelists as to the various aspects of social control. All of this genetic information—almost all of which provides for no cure and provides for no adequate prophylaxis of the predicted disease—is coming into a capitalistic, intensely competitive society where the divisions of wealth and power are growing wider day by day. These tests are coming into a society where electronic communication is making the dispersal of information instantaneous and where information in the hands of one institution can almost always be counted upon to be deliverable to another institution virtually at will.

This information is being generated by the combined medical, academic, and pharmaceutical industries. They are all in this in an interlocking way. They all have very serious financial interests at stake. Somewhere I read an estimate that this combined medical/academic/pharmaceutical industry is somewhere around 30% of our gross domestic product. They are the people who are developing this information.

To give you an idea of the value of the technology, last year I was in a seminar in San Francisco. A number of biotech companies were there, and the buzz at that conference was the fact that that week Rockefeller University, which had only two weeks previously announced the discovery of the *OB* gene thought to predispose some people to obesity, had sold its patent rights to a pharmaceutical company for—and this is the interesting number—an initial payment of $20 million and future payments, depending on the progress of the company in developing a test and/or treatment, of up to $100 million. That is just for one gene. The same story is being repeated almost daily as enormous sums are being expended for genetic discoveries that have not yet been fully exposed to independent peer review and long-term experimentation.

Most of us, geneticists will tell you, have a predisposition to at least one late-onset disease. The knowledge of who is likely to develop which disease and when is very intimate, powerful and dangerous. This information is being sought with various degrees of zeal by our largest institutions, the insurance companies, large employers and even our state and local government in a variety of different guises.

As many of you know, medical privacy in this country is virtually nonexistent. Licitly, or illicitly, any insurance company or other large institution can find out almost anything they want to about our medical history. While some steps are presently being taken to improve medical privacy, genetic information provides a special problem because, almost by definition, it involves entire families. Where one person is found to have a predisposition, frequently there is an inclination at least to test the other family members and that involves going out and talking to people and saying, hey, do you want this information? We are going to talk about insurance a fair amount today. I know several of our speakers are intimately familiar with some of the insurance issues. I had expected we were going to have a commissioner from the EEOC here to talk about genetic discrimination by employers. He is not here so I am going to talk about it just a little and then we will proceed.

There is a fundamental feeling that it is unfair for employers to use a genetic predisposition to deny somebody employment. Put yourself in the position of a woman who is being considered for a job promotion and who has a genetic predisposition to breast cancer. She is, by all current definitions, healthy. Yet, if, as some tell us, she has an 80% chance of getting breast cancer in the next 10 years, is she going to get that promotion once the company learns of her genetic status? I doubt it.

Or consider someone who wants to go into a management training program, but has a mutation thought to predispose him or her to early onset Alzheimer's disease. Despite the fact that that person will probably never get sick, his or her genetic status stands to ruin an entire career. In 1990, the ELSI folks made initial contact with the Equal Employment Opportunity Commission and the purpose of that was to inquire whether people with a genetic predisposition would be covered under the Americans with Disabilities Act. At that time, they received an opinion letter from Deputy Counsel for the EEOC that said that the ADA prohibits discrimination against people who *are* disabled or people who are *perceived to be* disabled, but does not protect people from discrimination who are likely to *become* disabled. There is a certain irony in that position, obviously.

More recently, the EEOC has reversed course and has published what are called guidelines for enforcement which do take the position that genetic discrimination by employers is illegal. Ultimately, this is an issue which will be determined by the courts. The courts are the only institutions in the society with the power to interpret the laws, and ultimately it will be the Supreme Court which will read the language of the ADA and decide how expansively to interpret it. Those of you with any familiarity of the law will recognize that over the last 20 years, the Supreme Court has been extremely conservative when reading and interpreting any civil rights legislation. Intuitively, to me, if you read the language of the ADA, genetic discrimination does not seem to be included. I do not think people with a genetic predisposition to a disease are going to be viewed as either having a disability, which is very closely defined within the ADA, or as likely to be perceived as having a disability even though they are not disabled.

There has been some tentative movement toward inclusion of genetic predisposition in state disabilities laws. To my knowledge, none of them have passed although some of them may have passed in this last session. (As these remarks are being supplemented, in September of 1997, no state has yet passed comprehensive employment genetic discrimination legislation, and several Congressional bills which would provide such protection are moribund.)

Finally, there is the issue of the social stigma which is attached to predisposition to disease. So let me raise some questions. What do you tell your fiancée? What does the adoption agency tell prospective adoptive parents if this information is available? Those are questions which have to be dealt with and, as this information comes increasingly into the flow of this society, I really hope the ELSI program and our legislators will get involved in all these matters.

Of course, the ultimate issue when it comes to genetic information and its misuse is that of eugenics. Bioethicists debate the meaning of this word endlessly. One thing is clear. Both sensible social policies and individual reproductive choices can easily verge into eugenics. As you consider what is and is not acceptable use of genetic information, I urge you to read carefully the words of Chief Justice Oliver Wendell Holmes in 1927. Many of the eugenic doctrines that reached their ultimate exposition in Nazi Germany had their origins in the U.S. By 1927 at least 20 U.S. states had passed laws calling for the involuntary sterilization of various "social undesirables." In the case of *Buck vs. Bell,* the Supreme Court was asked to rule upon the constitutionality of one typical statute. The person to be forcibly sterilized was a 16-year old girl with a questionable history of mental retardation in her family. In an infamous passage, Justice Holmes wrote:

> The judgment finds the facts that have been recited and that Carrie Buck "is the probable potential parent of socially inadequate offspring likewise afflicted, that she may be sexually sterilized without detriment to her general health and that her welfare and that of society will be promoted by her sterilization," and thereupon makes the order.

> In view of the general declarations of the legislature and the specific findings of the Court, obviously we cannot say as matter of law that the grounds do not exist, and if they exist they justify the result. We have seen more than once that the public welfare may call upon the best citizens for their lives. It would be strange if it could not call upon those who already sap the strength of the State for these lesser sacrifices, often not felt to be such by those concerned, in order to prevent our being swamped with incompetence. It is better for all the world if, instead of waiting to execute degenerate offspring for crime, or to let them starve for their imbecility, society can prevent those who are manifestly unfit from continuing their kind. The principle that sustains compulsory vaccination is broad enough to cover cutting the Fallopian tubes. Three generations of imbeciles are enough.

I respectfully submit that, if as great a thinker as Justice Holmes could succumb to the siren song of eugenics on the basis of the very limited knowledge of genetics in 1927, the danger that others will do so in the next 50 years as our genetic knowledge expands is great indeed.

# CAN WHAT YOU DON'T KNOW HURT YOU?: THE CASE OF GENETIC SCREENING FOR BREAST CANCER

Courtney S. Campbell

In the comic strip "Calvin and Hobbes," Calvin tells Hobbes as they are about to go on a sledding adventure, "I no longer wish to be called a 'boy.'" Hobbes quizzically inquires, "Isn't that what you are?" "Yes," Calvin replies while preparing the sled, "but I find that term demeaning and sexist." "What do you want to be called then?" Calvin shoves the sled down the hill and responds, "A 'chromosomally advantaged youth.'" Hobbes is skeptical: "That may not catch on," he concludes.

Knowledge of our genetic heritage and legacy can shape our sense of personal identity in fundamental ways, albeit comically distorted by Calvin. Nonetheless, one thing we have learned from research on human genetics and the Human Genome Project is that Hobbes' skepticism is warranted. None of us can be considered "chromosomally advantaged," for all persons are born with several defective genes. Yet, it is also clear that the genetic lottery has not dealt with us all equally. Besides the genetic mutations that cause disability, retardation, or even death at birth, we have learned in recent years that some persons are at greater risk for a disease that might afflict them in later life. The question is whether knowledge of genetic predispositions is beneficial, or whether the prospect of disease may evoke profound dis-ease in the lives of some individuals. What will be the impact of new information about our traits and susceptibilities on our sense of self-identity, and on our relations with others? I want to pursue this question in the context of discoveries over genetic susceptibility for hereditary breast cancer in women, using a story that I have been given permission to share.

## The Baconian Legacy

Contemporary medicine and medical ethics has flourished under the Baconian maxim, "knowledge is power." Much of the bioethics movement has

The author is an associate professor in the Department of Philosophy at Oregon State University, Corvallis, OR 97331.

117

devoted itself to rectifying power imbalances between self and nature, or between patient and professional, by providing more information about health and disease to the patient. Such information, including genetic information, is typically perceived as empowering moral agents, of giving them greater control over their lives, and their health care decisions.

In this context of unbridled optimism about medical progress, which will confer power and control to persons, the suggestion that some kinds of knowledge, such as information about genetic predispositions to disease, may be disempowering, seems to cut against this ethical legacy. Indeed, for some persons, this idea smacks of a different legacy, the paternalistic maxim that "what you don't know won't hurt you." Persons or advocacy groups that opt to resist genetic screening may thus be subject to criticism on two grounds; that they stand in the way of medical progress, and that their decision to live in ignorance may bring bliss but is not fully autonomous.

## A Case Story

A human face needs to be brought to this question. This story begins on one of those rare Oregon days when the sun is out. Over the last several years, my good friend (MGF) and I have deepened our friendship as we watched our daughters play on the same softball team. On one of these occasions, our conversation turned to counseling of persons with disease, and she off-handedly mentioned she fell into that situation herself. When I asked why, MGF related that recently she had been in for a routine physical examination, and her doctor had informed her that, given her family history, she should consider taking a new genetic test, the test that determines mutations on the *BRCA1* gene. I was aware from some of our past conversations about her family history that both her grandmother and her mother have succumbed to the disease and, moreover, MGF had described their dying process as "horrifying."

Why take the test now, when she is asymptomatic? Besides shedding some light on MGF's future health prospects, of course, her physician also mentioned a basic pocket-book issue. If the test was performed now, rather than waiting until symptoms develop, the physician believed she could better manage the flow of information so that the test results would be kept from the insurance company.

All this as we watched 20 young women try to do their best Dot Richardson imitations on the softball field. MGF had both of her daughters playing that day, a 14-year-old and an 11-year-old. She is both mother and very close friend to them, and I sensed much of her ambivalence about her choice revolved around her concern for them.

So, what was MGF going to do? The short answer is that she didn't really know. "What good would the test do me? ... I already know my family history. I know I'm at risk; I'm conscious of that in my life, and what I can do by way of risk reduction." Intimations of the complex relationship between genetic, embodied and personal identity surfaced: "If worse comes to worse, I can live without my breasts; I'm not sure how I would handle not having my ovaries, though."

As personal as this issue was, MGF was acutely aware that others within her family could be affected by her decision. There were her two sisters with the same genetic legacy, but the bonds of kinship were not particularly close between MGF and her sisters. Her daughters were a quite different matter. Too young to recall the death of their grandmother and just entering the roller-coaster of adolescence, MGF expressed a sense that her daughters needed to be more fully aware of her health situation—and potentially theirs—and also a concern for the vulnerability of their personal futures: "I can talk to them more directly about this, and what actions they can take to be healthy. But damn, they are too young to have to start worrying about this for the rest of their lives."

There was, however, one issue on which MGF was emphatic, and that was keeping the insurance company out of the picture, regardless of whether she took the test, or whether her results were negative or positive. She was quite convinced that nothing good would come from the insurance company becoming a player in her medical decisions. This suspicion is not without reason: In one recent case, a woman with a positive *BRCA1* test opted for a bilateral or modified radical mastectomy, citing personal reasons. Her insurance company refused to pay for the surgery, reminding her that "optional surgery" was not a covered expense. The woman's reticence stemmed from her fear that, upon disclosure, her insurance company would cancel her policy or raise her premiums. ("Patient encounters genetic discrimination," *The Oregonian* 2/6/97, B7).

## Genetics and the Imperative of Progress

It is over the insurance status of genetic information that the imperative of progress becomes prominent. The general argument runs along the following lines. Biomedicine is just beginning to uncover the insights into disease opened by genetic research. More research is needed to provide better diagnostic clarity and prognostic value, as well as to eventually offer the promise of effective genetic therapy. However, such progress is being impeded because of patient reticence, and patient reticence is partially if not primarily attributable to insurance issues.

It is perhaps not surprising that the imperative of progress surfaces as what is at stake for biomedicine in the debate over the ethical, legal, and social implication of genetics research. As Daniel Callahan has argued, the progressivist impulse is itself a result of the Baconian legacy connecting knowledge and power. The question that genetic screening sharply poses is who will be empowered through genetic knowledge, and for some, it is clearly not patients, or women at genetic risk for breast cancer. Progress seems to have become an end and justification unto itself, and patients mere means to manage in order to achieve that end. And so it seems to me that it is entirely understandable, perhaps even commendable, for MGF to question the benefits to her of the *BRCA1* test. The sentiments expressed by a reader's survey in a popular women's magazine regarding their desire to know their tendency to develop a genetic disease also become explicable: 44% answered yes only if the information would reduce their odds of developing the disease, while 22% responded no under any circumstances (Ann Parson, "Genetic Tests," *McCalls* September 1996, 68-70). How ironic that the rhetoric of progress marginalizes the interests of precisely those persons in our society the biomedical research enterprise has sought to cultivate under a mandate of inclusivity.

## Genetic Paternalism

Unlike some theorists, I do not believe that patient reticence is merely a manifestation of the medical paternalism of the past. Rather, the concerns of women faced with genetic predispositions is the impact of genetic knowledge on personal identity and the continuity of the self. The rationale of one woman, whose family history and situation is almost identical to that of MGF's, for deciding against taking the *BRCA1* test is suggestive of a richer and more complex self than the current bioethics vocabulary of autonomy permits: "If I'm tested and have the mutation, what are the options? I could have a total hysterectomy to try and avoid ovarian cancer as well as a radical mastectomy to remove both breasts. Yet there's always the chance that malignant tissue will get left behind in my abdomen or chest. Then I would have gone through two horrible surgeries for nothing." (Ann Parson, "Genetic Tests," *McCalls* September 1996, 68-70).

One senses that for this woman a genetic Damocles sword hangs over her. Her sense of self is at stake; the "I" following the surgical procedures would not be the same "I" prior to the procedures. Her refusal manifests a self that is willing to live with uncertainty and tolerate unknown risks. However, our standard accounts of autonomy and paternalism do not readily accommodate risk and uncertainty in a chronologically constituted self, because they are concerned with the capacity of a person to choose at a specific moment of crisis.

Thus, I think there are some very valid reasons for limiting genetic screening within research contexts. Such settings do, or should, provide pre- and post-test counseling. The ethical problem of offering genetic diagnosis without effective intervention is compounded when counseling is not available. This may occur when screening is incorporated as part of insurance or workplace requirements, or through uncontrolled environments like the market in test kits distributed by biotech companies.

The research context provides safeguards for both the welfare and the autonomy of persons that are especially significant to affirm in the rush of private companies to seek market advantages by offering genetic services to "consumers." The consumers' welfare cannot be paramount in an entrepreneurial enterprise designed to enhance stockholder investment. Nor is autonomy respected when pre-test counseling is offered through package inserts and post-test counseling is received through the anonymity of an 800 hot-line, or from a surgeon not trained in counseling.

The future of genetic testing in the hands of private biotechnology is illuminated by remarks from the two companies that have developed tests for *BRCA1* for the marketplace. According to Mark Skolnick, vice-president of research for Myriad Genetics, the priorities of commercialized testing are: "First, you save lives, especially with closer breast monitoring and prophylactic removal of the ovaries.... Then, by testing large numbers of women and collecting information on as many as are willing, we can fine-tune our knowledge." (Patricia Kahn, "Coming to Grips with Genes and Risks," *Science* 274: 25 Oct 1996, 496-498). In short, the entrepreneurial demand has reversed the standard priorities of medicine and science. The entrepreneurial ethic is, "Let's do the surgery first, then do our data collection later to determine whether the surgeries were needed or not."

The CEO of OnCorMed, meanwhile, has touted to *The New York Times* the financial benefits of widespread testing to the health care system. Genetic tests will work to cut medical costs because an insurance program will be able to "identify the [persons] at extreme risk [from hereditary cancers] and manage them differently from the others." (G. Kolata, "Tests to Assess Risks for Cancer Raising Questions," *New York Times* 3/27/95, A1, A9.) We have returned to the ethos of patients as entities to be managed, and managed differently. In this setting, I fail to see the promise of benefits for patients materialize. Different management may simply exacerbate the current stratification that we have in our current health care system.

## Is Genetic Knowledge Power?

I wish to conclude with three observations: (1) I have tried to suggest that, contrary to the Baconian legacy, knowledge can be disempowering in the context of genetic screening. Part of my concern is that patient welfare seems repeatedly to be subordinated to other concerns, whether research progress, insurance solvency, or private profit. The humanity of the patient is diminished by the ethos of "patient management," and persons are diminished by the designation of "patient" when they are healthy.

A more adequate ethic for genetics will forsake the language of autonomy for the concept of care. The ownership model of medical information—which is closely intertwined with the ethos of patient management—is simply the wrong way to ask the question. Rather than ask, "who owns genetic information," we need to start with "who cares for the patient?"

(2) Little good will come, I believe, of making genetic testing the domain of private biotechnological firms whose interests lie elsewhere than those of consumer welfare. The "consumerist" model is fundamentally flawed, as a person has no control over their genotype in the way the rational consumer model

of decision making requires. A research model that views the subject as a "partner" in a collaborative project will offer us better data, the hope for more effective interventions, and a context of care.

(3) The moral villain in much bioethics literature on genetics is the insurance industry. Researchers find the lack of insurance coverage a hindrance to progress, while patients and their advocates, as well as physicians, may conceal diagnoses or tests from insurers. The situation seems designed for confusion and moral chaos.

There seems to be little momentum for the kind of systematic change in health care reform we really need. Recent health care reforms are important but insufficient. However, if genetics research continues at its current pace, we may all be subject to insurance exclusions or higher rates due to genetic pre-dispositions. Perhaps out of shared vulnerability will arise the motivation and personal involvement necessary to make the inequity and stratification of the health care system a matter again of public discussion. That is a hope, to paraphrase Hobbes, that I hope will "catch on."

# The Ethical, Legal, and Social Implications Program at the National Human Genome Research Institute, NIH

## Elizabeth Thomson

Today I will focus on the Ethical, Legal, and Social Implications (ELSI) program at the National Human Genome Research Institute (NHGRI—formerly the National Center for Human Genome Research) at the National Institutes of Health (NIH). Much of the information that I am about to share came from a report called *ELSI 1990-95: A Review of the ELSI Research Program and Related Activities.*[1]

From the beginning the architects of the Human Genome Project realized that the discovery of new genetic information would have a significant impact on people's lives—on individuals, on families and on society. As a result, ELSI programs were established at both the NIH and at the DOE. These programs were viewed as necessary and integral to the success of the HGP. Originally, the NHGRI devoted 3% of its total funds to ELSI research and education programs and related activities. However, since its third year of existence, the ELSI commitment at NHGRI has exceeded 5% and it is anticipated that it will remain at that level or greater for the foreseeable future. This past year, this commitment of funds reached over $6 million in the regular ELSI research and education program. In addition to the $6 million spent in the regular ELSI program, another almost $1 million was spent funding ELSI activities within the specialized Genome Science and Technology Centers (GESTECS) and another $1.5 million was contributed to ELSI special initiatives by other NIH Institutes.

The original charge to the ELSI program was very broad—to identify, analyze, anticipate and address the ethical, legal and social implications of the Human Genome Project. This was to be achieved by: examining the ethical, legal and social consequences of this new information; stimulating public discussion (in fora such as these); and facilitating the development of sound policy options in order to attempt to assure that the information discovered is used to benefit

The author is Program Director of Clinical Genetics Research, Ethical, Legal, Social Implications (ELSI) Research Program, National Human Genome Research Institute, National Institutes of Health, Bethesda, MD 20892. The presentation was updated to include 1997 developments.

individuals and society. In the original five year plan (which was revised about three years after it was established), the ELSI Program was charged with two specific goals to: (1) develop a program to help understand the ethical, legal and social implications surrounding the HGP; and (2) identify and define the major issues of concern and begin to develop policy options to deal with them.

The ELSI Program has been successful in reaching its first two goals. Programs that allow us to better understand the ELSI issues related to the HGP have been developed. To my knowledge, the NIH had never before had an ethical, legal and social implications program that was directly tied to a specific basic science research effort. It was then, and remains today, a rather unique undertaking; some still view it as an experiment whose successes and failures will be judged at some point in the future.

In 1989, the NIH/DOE Joint Working Group on Ethical, Legal, and Social Implications (ELSI working group) was established. (Several speakers at this meeting were involved in that original group: Pat King, Jon Beckwith, Bob Murray, and perhaps others.) The ELSI working group provided overall guidance in the early development of the ELSI programs. They were involved in the discussions which resulted in the establishment of research and education programs at both NIH and DOE. In 1990, the NHGRI extramural ELSI research program was established to support research and educational projects in institutions throughout the United States. The ELSI working group assisted in framing the original research agenda for the ELSI program. With input from the working group, the extramural research staff developed and released the first program announcement later that year. This program announcement cast a very broad net, in terms of the kinds of research that the program wanted to encourage and support. It highlighted the need for research regarding fairness in the use of genetic information; the impact of genetic information on people; privacy, confidentiality, ownership and control of genetic information; issues surrounding genetic counseling and genetic testing; and conceptual and philosophical implications raised by the HGP. Although initially the community of applicants was limited and the number of projects funded small, this program has now developed into a highly competitive and robust research program.

In 1995, the Office of Policy Coordination was established in the NHGRI Office of the Director. This Office was set up to help provide information and analysis on ELSI issues. In 1996, the intramural NHGRI program began establishing an ELSI program to support research within the intramural program and also provide resources to the intramural scientists at the NHGRI. Thus the ELSI program at NHGRI has grown substantially over the past six years.

As the ELSI program at NHGRI has matured, the level of funding has grown from $1.3 million in its first fiscal year (FY90) to $6.2² million in FY95. During its first year, much of what was funded were efforts aimed at bringing groups such as this together to discuss and develop the research agenda for the ELSI program. We are now finishing FY96 and the budget for this year will be about just over $7^2$ million of NHGRI funds. In addition, about another $2 million could be added to that funding base because other institutes at the NIH have jointly funded some of ELSI's special research initiatives. The National Cancer Institute, The National Institute of Mental Health, The National Institute of Nursing Research and the National Institute of Child Health and Human Development have all co-funded research projects that were funded in response to one of ELSI's special initiatives.

The ELSI program has established extensive collaborative and co-funding relationships with other components of the NIH community and also other federal agencies. When the HGP was first begun at NIH, there was limited interest in the ELSI program. No other Institute or Center requested that they be identified as interested in funding such research by signing on to the initial program announcement. However, during the last year, when the ELSI program announcement was revised and it went through the NIH system, a number of Institutes responded and requested clarification about our intentions. Some wanted to assure that we were not overlapping with their interests. Others expressed an interest in supporting work in this area as well, and some considered signing on. In the end, two Institutes formally signed on to the ELSI program announcement. They were the National Institute of Mental Health and the National Institute of Nursing Research. The National Cancer Institute has also formally expressed their interest in these issues. In their 1997/98 budget proposal,[3] three specific goals offering new opportunities in cancer genetics research were identified. The first goal to be achieved within the next five years is to identify every major gene that predisposes people to cancer. The second is to use this information to transform medical practice. The third is to identify and solve psychosocial, ethical, and legal issues associated with cancer genetics. These are ambitious goals, and it is likely that they will have a challenge ahead to meet them. Finally, in addition to collaborations with other NIH components, the ELSI program has collaborated with a number of other federal agencies, including the Department of Energy, the National Centers for Disease Control (CDC) and Prevention, the Agency for Health Care Policy and Research, and others. As a result of the development of these important relationships, the ELSI program has been able to be more effective in dealing with issues of mutual interest and concern.

Since the ELSI program began, four major priority areas have emerged. The first involves issues surrounding genetic research. This includes issues of informed consent, privacy, commercialization, patenting, and other issues related to genetics research. The second priority area is related to the clinical integration of new genetic technologies. It should not be surprising that the NIH would identify issues related to human health and health care and technologies as one of its highest priority issues. The third major priority area is privacy and fairness in the use and interpretation of genetic information. This includes not only genetic privacy and genetic discrimination, but also examining the philosophical assumptions, the theoretical underpinnings, and the conceptual framework for how people think about, interpret, and use genetic information and knowledge. Finally, the fourth priority area is genetics education for health professionals, other professionals, and the public, with education of health professionals having highest priority at NHGRI at present.

A couple of specific examples will be illustrative of NHGRI's ongoing ELSI activities. Genetics research has brought into focus some new issues of concern related to the informed consent process and the protection of human participants in research. Participation in genetics research can result in the discovery of powerful, private, and potentially predictive information. It can result in the disclosure of unanticipated or undesired information, or information that has familial implications. The process of informed consent was designed to ensure that people are told about the risks and benefits of participation in research, of having a test, of having a surgical procedure, of taking an experimental drug or receiving an experimental treatment. Historically, a good number of the risks about which people have been concerned are physical—the risk that a complication will occur or that harm will come, or even that death might result. The major risks associated with participation in genetics research are psychosocial rather than physical in nature. This shift in focus from the limited physical risks associated with blood drawing to the potentially much more significant psychosocial risks has led to what some have called a paradigm shift in the concept of informed consent.

With regard to genetics research, the major risks are associated with getting information that may cause anxiety, may have an impact on family relationships, and that may result in stigmatization and discrimination. Because for years, genetics research involved a "blood draw" only, almost all such proposals which involved human participants were reviewed by Institutional Review Boards (IRBs) using the so-called "minimal risk" review criteria. Such a review has been commonly found to be inadequate to seriously consider the issues mentioned above. As a result, the NIH peer-review group for ELSI research proposals (which involved genetic testing as a part of the research projects) were often found to have been insufficiently reviewed by local IRBs and human subjects concerns were commonly raised as a part of the review. This observation has been communicated to investigators and also to research institutions throughout the U.S. and now most such research proposals are reviewed with a far more thorough evaluation.

The HGP has also funded research projects in which stored tissues were to be used for genetics research. Again, many such proposals had been approved without a comprehensive review by the IRB. In fact, many such proposals had an expedited review, or in some cases, it may not have been reviewed at all because of the assumption that the research did not involve human subjects at all. At NHGRI, we found ourselves repeatedly asking, "From where did those stored tissues come?" The tissues which had come from human research participants, in some cases, had been shared among researchers and institutions. Often, the tissues were still linked to an individual's name or other identifiable information. At times the informed consent had been less than optimal. This resulted in our becoming increasingly concerned that genetic information could be discovered about and possibly disclosed to individuals who may or may not have agreed to participation in the proposed research. As a result, NHGRI co-sponsored a workshop with the CDC to discuss issues surrounding informed consent for genetics research using stored tissues. This workshop resulted in widespread deliberations on this topic and the publication of a paper in JAMA (Clayton et al., 1995). In it, recommendations were made about what issues should be taken into consideration by an IRB, researchers, and others when stored tissue samples are to be used for genetics research. As a result of this meeting and the subsequent deliberations, there was a flurry of activity surrounding this issue. The American College of Medical Genetics (1995) has published a statement on this issue, as has the American Society of Human Genetics (1996). The College of American Pathologists and other pathology organizations (1997) have produced two documents on this topic as well. Most recently, the President's National Bioethics Advisory Commission has begun examining this issue as well. As a result of all of these deliberations, there will likely be some further clarification of the regulations regarding the use of stored tissue samples for research purposes.

One more recent specific and related issue of concern involves the use of stored tissues for the large-scale DNA sequencing associated with the Human Genome Project. In 1996, the NHGRI funded a number of pilot projects to begin the large-scale sequencing of human DNA as part of the HGP. At the time these sequencing projects were being reviewed, it became clear that the DNA libraries that had been constructed for use were being constructed from a small number of individuals, rather than the large number of individuals that had been previously assumed. In some cases, the individuals who had donated their DNA for the construction of those libraries were identifiable by someone in the laboratory constructing the library. Further, it became clear that the process of informed consent for some of the libraries had not been optimal.

A review of the federal regulations that guide the protection of human subjects in research[4] made us aware that in large-scale DNA sequencing projects, it might be difficult to completely adhere to the strictest interpretation of all of the existing regulations. Since the federal regulations are open to some interpretation, it was necessary for NHGRI to attempt to examine these issues and seek their further clarification. For example, the federal regulations require that research subjects "...may discontinue participation at any time without penalty or loss of benefit to which the subject is otherwise entitled." In large-scale sequencing projects, the libraries that are constructed for DNA sequencing will be made publicly accessible and widely distributed throughout the world. Once the libraries have been constructed and distributed, withdrawal from participation may be difficult, if not impossible. Thus, there is only a limited time in which research participants can withdraw their research participation. A second example is that federal regulations require that there be "adequate provisions to protect the privacy of subjects and confidentiality of the data." In large-scale sequencing projects the sequence information will be placed in a public database. The information obtained through large-scale sequencing efforts will not be kept confidential and cannot be removed from the database once it has been placed in the public domain.

After months of deliberations, NHGRI, along with DOE, developed a policy regarding human subjects issues related to large-scale DNA sequencing. It states that the best way to protect DNA donors to the Human Genome Project was to ensure: that the initial version of the complete human DNA sequence is derived from multiple donors from diverse populations; that donors would have the opportunity to make an informed decision about whether they would want to contribute their DNA to this project; and that library constructors take effective steps to ensure the privacy and confidentiality of donors. This policy is now in effect and further clarification of this guidance is still occurring. The policy discusses the importance of identifying volunteers who are willing to donate their DNA for the Human Genome Project. It specifically discusses the information that needs to be shared in the informed consent process and suggests that special care must be taken to protect donor identity in order that individual donors will not be identified and their genetic makeup become known.

A second area of emphasis in NHGRI's ELSI program has been in the area of clinical integration of new genetic technologies. NHGRI has funded a series of projects to examine the implications of genetic testing for cystic fibrosis in large populations. In addition, a set of projects designed to examine the use of genetic tests to identify those at increased risk to develop cancer has also been funded. In each case, a research consortium was formed to promote collaboration among individual investigators, reduce duplication of effort, increase the likelihood of interstudy comparisons, jointly develop guidance for informed consent and, in the case of the cancer genetics studies, develop follow-up recommendations for those found to have breast, ovarian, or colon cancer mutations.

Both of these special initiatives resulted in interdisciplinary work that brought together scientists who were involved in the development and use of the technologies with consumers and scientists who were equipped to examine the psychosocial impact of using these technologies on people's lives. These initiatives have allowed us to gain empirical information about what happens to people in a more controlled research setting, before the technologies are introduced into mainstream clinical practice. These studies have examined knowledge and attitudes about, interest and demand for, and the impact of using new genetic technologies. They have also allowed us to examine how best to educate people about these tests and also how to get adequate informed consent for testing. Finally, they have allowed for a more careful look at what happens to individuals who are tested and in some cases those who are not.

Through these initiatives, we have begun to learn a number of things. One is that there are almost always unanticipated consequences of gaining genetic information through participation in genetics research or by having a genetic test. For example, most would not have expected to find people who had two CF mutations who clinically do not have cystic fibrosis (at least as it has been described for all these years). This resulted in the identification of individuals who had no clinical symptoms of CF and yet were expected to have them or get them at some time in the future. There were also situations in which, because the parents were both found to be carriers of cystic fibrosis mutations, they wanted to have their living, apparently healthy, children tested to see if they had cystic fibrosis mutations. They were concerned that their children indeed had CF and had yet to be diagnosed. This resulted in genetic testing on healthy children for a genetic disorder for which presymptomatic treatment had not yet been shown to be effective. In general, most believe that carrier testing should not be done in individuals younger than 18 years (ASHG/ACMG Report, 1995). However, such programs may result in the identification of children who are carriers of CF mutations, without them having the opportunity to make the choice about learning this information for themselves.

Some people may experience increased anxiety (albeit sometimes temporary) as a result of learning about genetic information, while others may be relieved. Gaining such information may result in altered family relationships. We are also aware that some may be stigmatized by the information and potentially lose employment or health, disability, or life insurance coverage. Those are some of the most significant risks about which we must be concerned at this point in time. However, we have to be honest about the fact that there may be some risks that we have not yet identified or predicted. Thus, building in an opportunity for recontact at the time informed consent is obtained is important in most cases. That means that people must be asked if they would give their permission for recontact should new information become available or further risks be identified. Since some people do not desire any further contact, they should have the option to decline.

As a result of the establishment of the Cancer Genetics Studies Consortium, three important papers have been published in JAMA and a fourth paper has been accepted for publication in the near future. In their early discussions about their projects, these investigators realized that, during the course of their projects, they would identify people who have breast, ovarian, and colon cancer mutations. It became clear to each of them that they would have to develop policies and recommendations regarding informed consent and also regarding follow-up for those identified to have mutations. Rather than each of them individually identifying their own protocols, they decided that they would attempt to develop joint guidance about informed consent and also common protocols for follow-up care, based on an evidenced-based approach. The first two papers (Burke et al., 1997a, 1997b), published on March 19 and 26 in JAMA, summarize the consensus recommendations reached by this group of experts regarding follow-up care for those found to have breast, ovarian, and colon cancer mutations. The third article on informed consent for cancer genetic testing was published in May. The final article scheduled to be published late in 1997 will discuss ethical and health policy issues in cancer genetic testing. There are numerous other examples of projects that ELSI is funding that could be discussed. Some are in the above identified areas, while others fall into the area of privacy and fair use of genetic information and genetics education for professionals and the public. In fiscal year 1997, the NHGRI ELSI program will be spending somewhat over $7 million in an attempt to further respond to the identified issues of concern. We look forward to future opportunities to share our goals and accomplishments in the years to come.

## Footnotes

1. Available at www.nhgri.nih.gov
2. Includes estimated funding for ELSI activities in GESTECS.
3. Bypass Budget Request. The National Cancer Institute, "The Nation's Investment in Cancer Research: A Budget Proposal for Fiscal Years 1997/98," May, 1996.
4. Title 45, Code of Federal Regulations, Part 46, Revised June 18, 1991.
5. NHGRI policy regarding large-scale DNA sequencing available at www.nhgri.nih. gov

## References

ACMG Statement. 1995. Statement on storage and use of genetic materials. American College of Medical Genetics Storage of Genetic Materials Committee. Am. J. Hum. Genet. 57:1499-1500.

ASHG Report. 1996. Statement on informed consent for genetic research. Am. J. Hum. Genet. 59:471-474.

ASHG/ACMG Report. 1995. Points to consider: Ethical, legal, and psychosocial implications of genetic testing in children and adolescents. Am. J. Hum. Genet. 57:1223-1241.

Burke, W., G. Petersen, P. Lynch, et al. 1997a. Recommendations for follow-up care of individuals with an inherited predisposition to cancer. Hereditary nonpolyposis colon cancer. JAMA 277(11):915-19 (March 19).

Burke, W., M. Daly, J. Garber, et al. 1997b. Recommendations for follow-up care of individuals with an inherited predisposition to cancer: *BRCA1* and *BRCA2*. JAMA 277(12):997-1003 (March 26).

Clayton, E., K. Steinberg, M. Khoury, E. Thomson, et al. Dec. 13, 1995. Consensus Statement: Informed Consent for Genetic Research on Stored Tissue Samples. JAMA 274:22.

Geller, G., J. Botkin, M. Green, et al. 1997. Genetic testing for susceptibility to adult-onset cancer: The process of informed consent. JAMA 277(18):1467-1477.

Wilford, B., K. Rothenberg, E. Thomson, and C. Lerman. Ethical and health policy issues in cancer genetic testing. Journal of Law, Medicine & Ethics. (In press)

# TECHNOCRACY AND DEMOCRACY

## Philip Bereano

The title of my presentation is *Technocracy and Democracy*. Democracy of course is rule by the people, and technocracy literally means rule by the technical elite.

The field of genetics is one in which more traditional notions of science and technology are really collapsing in upon themselves because you have such rapid attempts to implement or to utilize relatively new discoveries that we do not have the kind of lag or differentiation in terms of institutional infrastructure. So I will just use the term technology. Whether one considers the Human Genome Project science or technology, I want to lay out the position that technology, for want of a better term, is not neutral. Although the dominant ideology in this society claims that science and technology are neutral activities, I want to maintain that they are not.

What is technology? If you went out and asked people on the street, you would get examples that are very thing-oriented—like computers or automobiles. Certainly technology does encompass things, even things a little less tangible like credit cards and financial systems. There are some people who would say the insurance system is a form of technology. I think a broad definition of technology is most helpful for understanding the social activity or concept of technology. That is, that technology is not only the things and the processes involved in them, but it also involves the relevant institutions. In other words, it is not: here is technology and here is the social part. But, in fact, technology is a social phenomenon—the making of things, the application of things, the application of useful knowledge, bringing together those kinds of skills. Applying and utilizing them is in fact what needs to be the focus of attention. So rather than set off science and technology in one little box and social phenomenon in another, I maintain that they are entwined—as I titled a book, *Technology As A Social and Political Phenomenon* (1976).

The author is a professor in the Department of Technical Communication at the University of Washington, Seattle, WA 98195-2195 and serves on the Board of the Council for Responsible Genetics.
©May not be reproduced in any form without the expressed written consent of the author.

## Paradigms

Let me suggest that one might consider the relationship between technical and social activities in a number of different ways. There is a classical paradigm that used to be dominant in this country in which technology is equated with progress. In fact, up until the post World War II era, there were numerous examples of statements that basically connect science and/or technology with the notion of progress. Progress, by the way, is of course just a statement about means rather than ends. It does not tell you progress in terms of what. This is also true of technology as a tool or instrument; it does not tell you for what. But earlier in the century people did not ask those kinds of questions very much, and the notion that technology was equal to progress meant that technology was good.

It is possible to say that, other than for a few cranks and misfits, that was basically a pretty widely shared point of view until people like Rachel Carson and Ralph Nader and others came along in the early '60s to begin to articulate what larger numbers of people were beginning to know from their own experiences. That is, there were some problems with this cornucopia of technological goodies that was being churned out by the miracle of American science and technology.

As a result of what was called "externalities" by economists, things like pollution and things which are usually unintended and certainly side effects, different conceptions of the notion of the social definition of technology have been put forward. There are very few people now who basically will maintain that technology is equal to progress. There are, however, a few of them around. They are like museum pieces in a way—a little bit antique.

But generally, if you ask a lot of people about this today, they will fall into one of three general camps. The most dominant one in this culture currently, and one which many persons attending a Human Genome Conference would embody, is a kind of view that the British commentator David Dickson (1974) called the "use/abuse model of technology." In other words, technology is neutral. It represents basically neutral factors and can be either used or abused. It is all up to human beings. I do not know if this needs any more elaboration because we are being surrounded by that view all the time. This is what any dominant ideology does. It surrounds you with that point of view.

A second category is actually a grab-bag of a couple of categories. I apologize to people who might be partisan in some of the sub-categories if you feel it is an unfortunate lumping together. But I will do that just because I want to move ahead. It is sort of either an anti-technological view or greening view or the appropriate technology point of view. Basically, this view says there is something very wrong with the dominant technological systems, the social and political aspects of technology. They say either we have got to back off, we need less and we have to somehow step back and re-configure this very, very differently.

The last view, which I personally hold and which I want to elaborate on a bit more, is what I will call the "social relations model of technology." It holds basically that technologies are developed in a society such as this one—which is a class society—reflect the underlying relationships of power since the powerful sectors in the society are able to articulate the research agendas, get the funding and the programs going, have the science performed and then have the technologies developed. There is no better example of this than the Human Genome Project—the example I always use.

I do not know how many people are really familiar with some of the early stories about how the Project got going. I only know them second, third, fourth-hand. For example, some of the earliest opponents of the Human Genome Project were other biologists, who were afraid that there would be a diversion of resources at a congressional level to this new activity, and it was not until the molecular biologists basically got together with them and said: "No, let's go in for more money, new money, and so forth," and they had the power to do it.

We are talking resource allocation. We are talking about groups that are powerful enough to lay claim on a couple of billion dollars at a time when—well, as the newspapers headlined last week—"*$45 Billion Being Taken Away from Poor Families*" and things like that. You can use whatever examples you want. We are talking about the power to get your interest, your ideas, your agenda attended to. Incidentally, this has nothing to do with intrinsic worth or non-intrinsic worth. The issue here is political power and the power to use certain ideological configurations to sell it, to get members of Congress who know nothing about biology interested in and excited about the work you want to do, or whatever the technology is, or the science.

I want us to focus in on power. That is what this talk is about. I will use the term a lot, because I think it is at the root of a number of things that have been talked about today, a great group of presentations in terms of the variety and richness of where the viewpoints originated. I am not sure if anyone actually used the word "power" or not; maybe they did and I just did not catch it. But I want to talk about some of the elements of power that I think you will see were, in fact, reflected a lot in the talks that were given today.

To do that, and to make my point a little more concretely, another example that I think the social relations model of technology is the only one that really helps us understand is how technologies such as nuclear power get developed (and get special laws that shield it from liability, like the Price-Anderson Law) when other forms of power, such as solar power, go withering. This model helps us understand how patent law gets changed to accommodate the new genetic engineering even after 220 years of it being understood by everyone—by the way, I studied patent law and worked in a patent law firm—that living organisms could not be patented. That is why you had to have special legislation for plants, the Plant Patent Acts. All of a sudden, by one vote in the Supreme Court, the law gets changed, and with no further congressional action, no further court action, the Patent Office becomes the reflector and the propagator of the ideology that says the development of this technology is necessary for the economic well-being of this country. Therefore, that requires the patentability of this activity and a whole host of things.

Even at the time of the Chakrabarti case upholding the patenting of genetically-engineered organisms, I think you could hardly have found a patent attorney in the world who would have believed that human genes were patentable, or that a whole genome like that of the Hagahai of New Guinea would be patentable, and so forth. Now we are seeing the organs of the society facilitate this. This is what power brings about in my view.

To better understand it, I think we need to step back critically to look at the dominant ideology and how it facilitates these sorts of things. In my classes, I usually call it "corporate liberalism," not because it is liberalism as opposed to a Newt Gingrich kind of conservatism or whatever—a lot of conservatives in fact embody corporate liberalism—but because it grew out of a form of liberalism, as I understand that factor. I will give you several general principles—and you will recognize them all—but it is a way of trying to deconstruct the general notion of our dominant ideology. You know, the fish is in the water but the fish doesn't know it is in the water. Since we are so immersed in this ideology, sometimes it is helpful to try and step back and think about what this is all about.

**Corporate Liberalism**

The first kind of principle—and I am using it in regard to technological phenomena—is that *technology increases human options and hence human freedom*. The definition of human freedom is having more options to choose from, which is of course a market-based theory, consistent with capitalist ideology. Now the reality is that technology invariably can close off options, but that is not really talked about very much. It is very hard to buy an electric car or a steam-powered automobile, but those were real technological options 90 years ago. But technology is about institutions fighting for power. In that example, the institutions including Rockefeller at Standard Oil and so forth were able to assure the dominance of the internal combustion engine and to have the other forms of automotive propulsion atrophy as a result. But the dominant ideology just teaches us that: Technology increases human options and hence human freedom. By the way, if you just check out an elementary, junior high and high school curricula, insofar as they deal with any of this, you will see this ideology really reflected there. It is a little frightening from my point of view.

The second principle, as I began to state earlier, is that—*technology is neutral, objective and value free*—except for externalities and things which, by the way, can often be corrected by the use of more technology. We are worried about pollution. The answer to pollution is not to change our social ways, not to change, for example, our transportation system, but let us develop catalytic converters. I am not against catalytic converters, I am just against pollution. I used to work in the federal air pollution agency. I am just saying that this is the approach the ideology fosters: that the solution to technological problems will be technological. By the way, we are hearing a lot about that in regard to human genetic engineering. Rather than the majority accommodating to people, we are going to develop technological fixes. I was not here but the discussion of "the perfect baby" must have given many examples of this.

In my observation, the third principle is that—*no theory of social change, social causation or social reality is necessary in order to be an expert at discussing these things.* If you look at the dominant ideology, where they will say, for example, we have to support this industry for our economic well-being or whatever, there is never any theoretical construct laid out about how social change occurs. All the things having to do with relationships—you can pick anything, even non-technological things like increase in crime, decrease in crime—no one in this society feels obligated to start or reference or frame a discussion of these sorts of things in any kind of theoretical construct. I am talking in terms of the media and people talking on the streets about this. So when you hear all of these claims about technology, there is never any reference to the speaker's belief about how technological change relates to social change, or how social change comes about, etc. It is just: this will be good.

Fourth is that—*the relationship between social change and technology is usually presented as being unidirectional.* You may remember a television series a number of years ago called "Connections," which purported to show how all kinds of bizarre results flow from a single cause. This view presents social processes as a kind of one way street. I actually think it is obvious, if one really looks at it, that what you have is an interaction. It is much more dialectical. Social processes shape and form new technologies in just the way that technologies shape and influence social realities. The reason that this is presented as one way is that this is a very good shield for groups in power; it suggests that the technologies are inevitable.

Again, the Human Genome Project is about one of the very best examples one could use. By the way, I must say to the people here who have put lots of good years, time and energy into ELSI, that I think to some extent—I was asked to be provocative—ELSI serves as a cover for the Human Genome Project in exactly this way. It says, "Yes, indeed, there may be some problems coming out of this new technology. Let's see how we can address them," rather than dealing with, or suggesting that the public at large deal with, how this new science and technology got going to begin with? The problems and the relations that we are dealing with now have to be understood as a kind of dialectic in which powerful biological and economic interests, including venture capitalists, saw situations to push for what they wanted. They did not care about genetic discrimination. They did not care about problems of definitions of race that Professor Jackson and others talked about this morning. What they cared about was making a buck, and they saw a very good way. I am not saying that there is anything wrong with making a buck; I make a buck too. But what I am saying is I try to do it in a more dialectically conscious manner.

So we have this principle operating with almost all technological phenomena in this society because this is what the dominant ideology does. It disconnects the social, political and economic realities which create science and technologies and say: "Oh, if there is a problem, you start with the technology and then we see if there are any problems and how we can deal with them" rather than dealing with an organic, living, breathing interactive system, because this is a nice shield for the groups in power that are behind this. It is a disconnect in terms of following the path of power.

Fifth, most of the discussion about technology and social impact analysis—certainly technological impact analysis—deals with society either as an undifferentiated whole or else looks at the level of the individual, rather than dealing more appropriately with social groups and more organic structures that are intermediate in size. Again, I am talking more about a very general literature—the discussions and discourse that are out there. In meetings such as this, in fact, and that is one of the things that I really enjoyed this morning, there was focus on different levels, for example, racial or ethnic groups and whether those categories have any meaning, what they might mean, and how they relate to this kind of phenomenon. But basically, most of the discussions out there say: "This is good for society." Society is always discussed by a lot of these speakers as if it were some kind of independent organic actor with one mind, one kind of set of interests and so forth.

At the other extreme, we talk about individuals: "Oh, you are being discriminated against. Your genetic information might be known. What should you do about it?" Or: "You are using up too much water or electricity. You should drive less, or you should heat your house less." All these things are either some kind of grand social scheme which no one can get a hold of because how can any individual, especially non-elite individuals, affect society? Or they are devolved into highly atomistic, highly individualistic kinds of problems, in which it suggests to people that you are isolated in your alienation and your frustrations. Everybody else is able to manage two jobs and take care of the kids and everything. You must be a failure or are having some problems, rather than the problem being structural and organic.

What this does in much of the literature that analyzes technological impacts and tries to deal with the social reality of them is to fail to disaggregate. When talking about benefits, the better analyses acknowledge that there are risks and costs, as well as benefits. They don't just talk about benefits. But they fail even there, many of them, to disaggregate into these kinds of intermediate social groupings and to realize that very often the benefits fall on certain groups in society, and the risks and costs fall on other groups in society. To my mind, this is very, very close to the definition of power and the ability to set up a situation where you and your buddies benefit and other people pay the tabs.

We see this all the time with technological phenomenon, particularly among communities. For example, indigenous communities all around the world whose tundra is being torn up, whose resources, whether it is the people in Nigeria where there were the recent hangings because of the oil company exploitation and attempts to oppose that. Whether it be—someone referred earlier to the Human Genome Diversity Project—the belief among most indigenous peoples that, after mining their earth and taking all of their other resources, now people are going to come from the First World and seek the very people themselves as resources to be tapped, examined, banked and maybe exploited. This is an example of the failure to examine the disaggregation of the costs, benefits, and risks and see that they fall on different people and how peoples react to that. This is a reality, whether you agree with it or not. This is how most indigenous people who are addressing this issue feel about it—whether it is true or not. I need to make that clear.

## Conclusion

So I will wrap up now. I think I have made my major points: that certain technologies, at least in some of their aspects—or someone could argue in almost all their aspects—are not neutral, because technologies are the result of human activity and purpose. They are intentional interventions into the environment that would not otherwise occur. They embody and manifest human purpose and intentionality and, as such, they embody ideas and goals. There is nothing neutral about this at all; the question is whose goals, what goals, what values are embodied and transformed.

Whose purposes or intentions? We are not all in this together. The Pogo cartoon of the early 1960s, that was very powerful in the environmental movement, was of course incorrect. It was put forward as: "I met the enemy and it was us" as a way to diffuse an analysis that would point the finger of responsibility for pollution to people whose decisions were really responsible for it. The decisions of a relatively small number of people—they were probably men, they were probably white, they probably could all sit around one table—for example, to develop the one-way non-returnable bottle and then spend billions of dollars marketing it to us to show us that this was *our* real desire, because we could then go to the beach, have a Pepsi and then just toss the bottle away and frolic in the sand without worrying about returning it to the supermarket or grocery store.

The non-returnable bottle is a slightly different technological phenomenon from the returnable bottle. I do not know if people notice this. Anyway, that change was made by a small number of people, each of whom realized that their corporation—Libby-Owens-Ford making the glass, Coca-Cola making the syrup—each of them could make a fraction of a cent more on, what is it, eight billion bottles a day, or is it not that large? But there was some extraordinary number of soda pop bottles that were out there. This is exactly what I am talking about. Those kinds of technological decisions were made to advance certain kinds of interests and, of course, those people who stepped on broken bottles on the beach or people who had to endure the aesthetic problem of the litter were the ones picking up, very literally, after those profiting from it.

That kind of a pattern is what we are talking about. How that relates to the Human Genome Project, I hope, is a subject of discussion. How did we make the choice that this is the most important biology problem facing us in the society? Also, is the attack on these genetic diseases worth the allocation of $5 billion when this country is about twentieth in the world in infant mortality, when for a fraction of the expenditure, we could be saving many more lives than we ever can realistically hope to save in certainly the near term from the Human Genome Project—if that is what it is about? I can do utilitarian calculus too; I am not so removed from reality to not be able to do that. Those sorts of questions about the relationship of power and how the Human Genome Project expresses and manifests power and *whose* power will, I hope, move more to the center of these kinds of meetings and discussions.

## References

Bereano, Philip. 1976. Technology As A Social and Political Phenomenon. John Wiley & Sons, New York.

Dickson, David. 1974. The Politics of Alternative Technology. Universe Books, New York.

# THE ETHICS OF PREDICTIVE GENETIC SCREENING: ARE THE BENEFITS WORTH THE RISKS?

## Robert Murray Jr., M.D.

Although I was asked to speak about predictive genetic testing, I will be talking about screening rather than testing. There is an important distinction between the two. Testing, in the medical sense, involves the evaluation of a patient or individual who presents with a particular problem or one or more symptoms or signs that indicate a reason to conclude there is a medical problem. An example might be the case of someone who is having symptoms of hypoglycemia which may be seen as an early phase of Type 2 diabetes. This would be a reason for doing a test of that person's blood sugar. In the case of a genetic condition, there may be a family history usually of a condition that might be a reason for testing a family member. Screening, on the other hand, involves the administration of a test to people who may be at risk because they belong to a specific population but no individual in the group is necessarily suspect for having a particular genetic trait. In some cases the definitions overlap, and it may be very difficult to distinguish them. The ethical principles that govern them are the same for either one, but the emphases and dilemmas differ.

Let me start with the questions that need to be raised when addressing a new screening or diagnostic test. When is a test ready for clinical application and who decides? One of the things we will learn is that in many cases there is no authority that actually makes it clear or decides when a test is acceptable. The Food and Drug Administration has traditionally done that, but many tests are introduced into medical practice without the imprimatur of the Food and Drug Administration or any other recognized medical authority. There is a difference of course between the clinical application of a genetic test in a given at-risk family and population screening. Because of the potential for making lots of money, there is a lot of pressure to introduce the wide-scale use of genetic tests. There is much pressure to go directly from the introduction of a *BRCA1*

The author is Chief of the Division of Medical Genetics in the Department of Pediatrics and Child Health and Professor of Pediatrics, Medicine and Genetics at the Howard University School of Medicine, Box 75, Washington, D.C. 20059.

genetic test to *BRCA1* screening of women considered to be at high risk. Other important points to consider are whether the population is ready or receptive and whether testing should be voluntary or mandatory. The public health agencies that promote testing are interested in not missing any possible cases. But, from an ethical standpoint, it is important that individuals determine whether they want to be tested especially when the benefits of such a test are not clear. This means that getting informed consent is critical.

Other questions to be considered are: who controls the results of the tests, and who should have access to those results? Traditionally, in medical practice, it is the patient who has access to the test and only the patient, and test results are not disseminated without their approval, or at least that is what it says on general consent documents. We know that often those tests are shared with other agencies and interested parties, and the patient may not be aware of it.

Can people with positive tests be protected from the harmful or unethical use of results of genetic tests by third parties? This question will be discussed in detail by other speakers, and I am not going to say a great deal about that although we have heard much about the interest of other parties in the results of tests including family members. The idea has been presented by some that genetic information does not belong to the individual but to the family.

There are vital ethical issues around prenatal testing for genetic disorders which will not manifest themselves until many years later in life. There are geneticists who have considered screening programs designed to reduce the burden of disease in the population by terminating the pregnancies of infants who show markers for such genetic diseases but who may have 30, 40, 50 years of life before those disorders are expressed. This goes along with what Dr. McGee had to say about the perfect baby syndrome. Is it a real possibility?

Finally, what makes genetic screening or testing different from testing and screening for other conditions? Many of my colleagues say there is no significant difference. It is just that genetic tests are more sophisticated and allow broader predictions extending to other family members.

Every other week it seems one sees in the paper or a magazine (which are often the primary sources of information on advances in genetics) a new gene of the week. The *Washington Post* recently carried a story reporting finding a gene that may be associated with prostate cancer. Furthermore, the implications of such genetic predictability are profound enough so that young people in high school are concerned with them. Young people learning about genetics and DNA testing realize that they may be marked, if you will, if they know that they carry a particular gene and that having an affected relative might stigmatize them. It is not just that they have the gene but, if their brother, sister, uncle, aunt or even a cousin has the gene, it means that they may be at increased risk.

A play was written by high school students with the support of NIH (National Institutes of Health) scientists in which the students dramatize some of their concerns around DNA genetic diagnosis. They had been learning about the power of DNA testing and sequencing. In this play they express their concern about an additional burden. Not only do they have to worry about the degradation of the planet, the ozone layer, and the possibility of a worldwide atomic holocaust. Now they also have to worry about the dangers of DNA testing as well. These young people are sensitive to issues such as what genetic predictability might mean for a given individual?

In a cartoon in one of the lay publications, a doctor is shown giving a report to his patient. The legend at the bottom says: "Great news, Mr. Jones! We found in your DNA that you'll likely have stomach cancer at age 52, colon cancer at age 58 and, if you're still alive, prostate cancer by 65." You might say, "Ridiculous." Not so. One's predisposition to cancer is a predictor of an increased risk to more cancer. If you get one cancer, you are more likely to get another cancer. That is one thing that people are generally not aware of and is something we learned without DNA testing.

So that is a possibility and the replicative mechanism of genes is such that errors regularly occur. Just think about the mice model that was presented yesterday having to do with protecting them from breast cancer with the addition of the *BRCA1* gene. One of the reasons why it only works for a while is because other errors are occurring in cells, and those errors are not tied in directly to the *BRCA1* gene. We may not be able to escape our fate. Nature did not make the replicative mechanism perfect; mistakes are made all the time. Remember that mutations are mutations whether they cause disease or not.

## Ethical Principles

That is the background of some of the concerns. We are aware of all of the good things that Dr. Botstein and others have told us about genetic testing. We are also concerned about the harms. That is why ELSI (Ethical, Legal, Social Implications Program) was established as part of the National Human Genome Research Institute at the National Institutes of Health by Jim Watson. I think he was very wise to do so although some of his colleagues did not agree. The main ethical values which I will speak of are: autonomy, beneficence as opposed to malevolence, justice, virtue and caring.

Autonomy is defined as self-determination, free of coercion; it is sometimes considered synonymous with freedom. This principle is tied to informed consent. The individual in this case controls the initiation of testing, taking a specimen, use of it, access to the results, and use of the specimen in current and future research. The principle of autonomy is protected to the utmost by the principle in medicine that says: first, do no harm. If you cannot be sure that you are not going to harm the patient, then do not do anything.

Beneficence has to do with not only doing good but preventing and removing harm or evil. In a sense, we try not only to promote the welfare of the patient by using DNA testing, but we also must try to prevent the harm that can come from such testing. We try to balance which is better. The question about "knowledge" is a big one because there is a belief in science held by many of us in the teaching profession that knowledge is a "good" no matter what. There is a lot of evidence now that knowledge is sometimes not a benefit, and that, on occasion, being ignorant is better than knowing from a practical standpoint.

So we must weigh goods vs. harms. Justice emphasizes the idea of fairness in the distribution of goods and benefits. It says that everybody who is tested is going to have access to whatever benefits may come from being tested. Yet, receiving benefits is tied into our health care and insurance systems, and we know that those may be inherently unfair in some ways. In this respect, we have a serious ethical problem.

Other values deriving from these principles include privacy, confidentiality, truth telling, personal and societal well-being. The conflict that may exist between the individual and society is serious. That is to say, some people argue that genetic testing should be done in families because that would bring the greatest benefit to the greatest number, and society would benefit. But whose society are we talking about? Are you talking about my perception of society as an African-American? Are you talking about my concept of society as a middle class American or just what? Or is one just talking about the United States of America in which case there is little doubt that U.S. society is *not* concerned about *my benefit*—at least not on balance and historically speaking. Useful knowledge is of course a derivative value. There are number of other values that we could consider. These are just some of them.

Some of the goods and harms related to health care—and I talk to you about genetic testing assuming it will take place in a health care setting—are elimination of disease, relief from suffering, amelioration of pain, improvement of handicapping conditions, and prolongation of life of good quality. Prolonged life in and of itself is not necessarily a benefit but, if it were of good quality, it would be. On the side of harm, there is of course illness, disease, needless pain and suffering, handicapping conditions, and premature death.

**Guidelines for Ethical Conduct**

Genetic testing and screening have been analyzed by a number of groups that were concerned with the ethical and social problems involved. Guidelines were developed. Back in the 70's, as a result of the sickle cell situation which I will mention later, a genetics task force of the Hastings Center recommended these five guidelines for any genetic testing program. Following them would make a program ethical.

(1) The community should be educated prior to the initiation of the testing program, and they should understand the benefits and harms.

(2) Informed consent and voluntary testing is preferable. That is one of the hardest ones to implement because the breadth and depth of knowledge of the population varies so much that giving informed consent is a problem. It has to be tailored to the individual.

(3) Accurate diagnosis is critical. Results should be unambiguous, and there is very little of a DNA testing that is completely unambiguous. They may give

you relative risk figures, but they are not absolute determinants of whether you will get a disease or manifest abnormality.

(4) Professional genetic counseling should be in place and accessible. That is something that is clearly not available. If we were to launch a massive genetic testing program across the country tomorrow, there would not be enough genetic counselors to meet the demand.

(5) Test results should be kept confidential. You will hear later of a serious problem here.

In 1983—about 10 years later, a President's Commission came up with a list of five ethical values as guidelines for genetic screening programs. They put confidentiality at the top of the list. The confidentiality of genetic information should be protected, and autonomy is preferred. Mandatory programs must always be justified. In other words, the burden of proof is on the person who says testing ought to be required in a given instance.

Knowledge should be disclosed where there is a need to know if other family members, for example, should be told. One has to justify whether or not one wants to reveal otherwise private information but, in the final analysis, it is my belief that confidentiality overrides the need to disclose. That is, privacy, in the case of a conflict, may be worth more.

All pre-screening and follow-up services should be available and pilot programs ought to be conducted before a screening program is initiated. Pilot programs for predictive genetic testing have been going on for Huntington's disease in Canada, and there may be one for this disorder at Johns-Hopkins University. Pilot programs for cystic fibrosis testing are currently underway which are being evaluated relative to benefits versus harms. But, in general, there have not been good pilot programs to evaluate screening programs. Finally, access to screening should be consistent with principles of justice. In fairness, I have already mentioned the problems of justice and access to health care; so we realize that there is a problem there.

## The Sickle Cell Debacle

It is apparent that failure to follow these guidelines causes problems. I am going to tell a story. It is about testing for sickle cell trait—not the disease but the carrier state for sickle cell disease. People who have the carrier trait are usually quite healthy. There is some evidence that a very small minority of individuals who are carriers of the sickle gene may be at risk for certain physical problems under certain extreme conditions. But these are very rare, and there is still no definite proof that the presence of the sickle cell trait is responsible for these abnormalities that have been reported in a small percentage of carriers. So we are not dealing with a disease like cancer or some of the behavioral disorders that cause serious disease.

In 1970, sickle cell disease was rediscovered in a sense by certain physicians who realized that we had a lot of knowledge about the sickle gene that was not being used. We knew its genetic structure. From its hemoglobin structure we knew a lot about its function. But nothing much was being done with the knowledge of this disease and its treatment. They insisted that we ought to pay more attention and do more research into treatment for this disease.

Secondly, the sickle cell trait was thought to be responsible for the sudden unexpected death of four army recruits during extreme exercise. This was reported in the *New England Journal of Medicine* and caused a tremendous furor about the possibility that the sickle cell trait was a dangerous condition. Of course, if this was the case, we would see thousands of Africans and West Africans dying because in some places 30 to 40 percent of the population may carry the sickle cell trait. But, of course, this was not happening, so it was obvious that the sickle cell trait was not primarily responsible for the deaths of those recruits to those of us who knew a little bit about the condition. Nevertheless, the National Academy of Science convened a special committee funded by the Department of Defense which was concerned about its African-American soldiers who might have the sickle cell trait and whether this was a dangerous condition. The finding of the committee, which I happened to chair, was that there was no clear association. But there were recruits with sickle cell disease getting into the armed forces through the medical screening. We suggested that they might do a sickle cell testing so that they would not recruit into the armed forces people with sickle cell disease.

President Nixon two years later recommended that $6 million be spent for sickle cell research. That was at a time when his administration was taking away about $50 million from Head Start programs. The $6 million for sickle cell was primarily a political move. At the same time, he recommended $25 or $30 million for cancer research because he could not leave out the rest of the population. The National Sickle Cell Anemia Control Act was enacted and African-Americans got very excited about those events. Finally, they had their own disease! People started doing sickle cell screening all over the country. A very simple precipitation test was introduced that required no laboratory skill whatsoever. A couple of drops of blood was placed in a solution. If the solution turned cloudy, the test was positive; if it did not turn cloudy, it was negative. The trouble was this test did not distinguish between sickle cell trait or the disease. So lots of people who had a positive test thought they had sickle cell anemia. This caused a tremendous amount of anxiety in the African-American community.

At the same time the NIH established some comprehensive sickle cell centers through the National Heart, Lung and Blood Institute to do research, and mass screening programs were established without sufficient education of the public. In fact, many people thought there was an epidemic of sickle cell disease because they never heard of it before. People in good health were thought to have this condition after a positive sickle cell test. Given the knowledge of people about genetic diseases at the time and given the fact that in some cases they were doing sickle cell tests on blood specimens taken to test for syphilis, some people thought that it was a sexually communicated disease. You can

144

imagine the confusion that reigned because the public was largely uneducated about the true significance of the tests.

Eventually, people learned that there was no safe prenatal diagnostic test available for a couple who both carried the gene. The mother and father must both be carriers. Such a couple at that time had no option to avoid having a child with the disease. If they were both carriers, the only way for them to avoid having a child with sickle cell disease was not to have kids. So some people concluded that this might be a genocidal program—a way of keeping black people from having babies.

More sickle cell centers were established, and a massive national education program was begun. But not before some states enacted mandatory testing laws for sickle cell trait because of the concern that small children entering school and playing at recess might start collapsing because of the supposed association with exercise. All or most of these laws were proposed by black legislators who thought that they were doing something positive for their black constituents. Fortunately, at the time the educational program was launched, some citizens launched legal suits to block such testing because it was only mandatory for blacks and not for whites. In some states, instead of the law saying that African-Americans were to be tested for sickle cell trait, they said everyone was to be tested for sickle cell trait except... then they listed all ethnic groups except blacks.

In 1975 other disease-oriented groups became aware that lots of money was being spent for sickle cell programs. Thalassemia, hemophilia, and various other special interest groups wanted their piece of the federal pie. An omnibus genetics bill was introduced which set model standards and a rationale for genetic screening programs.

There were other negative effects, at that time, that came from being identified as a sickle cell gene carrier. Psychological distress was created. Some kids began having episodes of chest pain and abdominal pains. When they had such, some mothers thought that their children were having a sickle cell crisis. The supposed connection between exercise and sudden death meant that black kids all over the country were required to be tested for sickle cell trait before they could go out for sports, and their parents had to sign waivers that, if anything happened to the child, the school was not to be held responsible and so forth. So this policy had the effect of excluding a lot of young people unnecessarily from participating in athletics. Some parents were truly fearful that their child might drop dead. Discriminatory laws that mandated sickle cell testing were carried out and the kids at school—especially elementary school—were stigmatized because it was very clear from the way the tests were conducted which kids had positive tests and which did not. The kids with positive tests took letters home to tell their parents that their child tested positive for sickle cell trait. The kids that tested negative did not get a letter. The kids that had sickle cell trait were vilified by their classmates.

About that time, Dr. James Neel, who is considered one of the fathers of human genetics and whose word carried a lot of weight, gave a paper in which he suggested that the life expectancy for an individual with sickle cell trait was 5% shorter than the average person without the trait. This led 27 percent of the insurance companies to increase the premiums on life insurance for persons with sickle cell trait. This increase was subsequently rescinded when other investigators did studies that did not support Dr. Neel's findings.

Selective screening of blacks in industrial settings led to some being excluded from some "high-risk" jobs. They were also excluded from being pilots, not only in commercial airlines but in the United States Air Force Academy. This limitation has been recently modified but not eliminated. Some black flight attendants were eliminated from their jobs even though they had been flying for years without any ill effects merely because they had the sickle cell trait. This is only a partial list of some of the problems that arose from poorly planned screening programs.

As testing became more widespread and physicians became more aware of sickle cell disease, they began to find this disease in white people, or people who said they were white. In any case, there were mysterious diseases that had gone undiagnosed or suddenly misdiagnosed because it was found that the sickle gene was more widespread in other ethnic groups than was generally recognized by the medical community in the U.S.

After several years, we began to get some resolution of the problems that had been created. In 1978, a way of diagnosing sickle cell anemia using RFLP methods and DNA was developed so we could offer a positive reproductive option to couples at risk identified by sickle cell screening. A study of the natural history was instituted to identify the problems caused by the disease. Selected states began newborn screening for sickle cell anemia. One or two, I think, began as early as 1976 but they did not fund them very well. In 1983, the first real breakthrough in screening came about when prophylactic penicillin was found to be a means of reducing mortality and morbidity in sickle cell anemic infants. This was officially accepted in 1987 by a Consensus Development Conference at NIH. Now, newborn screening is the major focus of screening for sickle cell disease. We are finally screening in a way that we can do something about managing the disease.

If we were screening in the past primarily to provide treatment for disease, what are the reasons for screening now for individuals at increased risk who do not have a disease to be treated? Well, on the pro side or the benefit side, we want to reduce morbidity and mortality of disease. The idea is, if we find the disease early, we can treat it. Early diagnosis and early treatment means the possibility of a life saved and morbidity averted. This is not always true, unfortunately. In fact, it is true for only a minority of conditions.

Early identification could avoid much pain and suffering if we are able to modify the course of the disease. We can increase the life span and maybe provide a better quality of life.

Screening will identify members of the high-risk population and therefore we can focus our attention on people who are at increased risk. Using what we call susceptibility genes we may not be able to diagnose the disease to which they are susceptible, but we can focus our attention on the patient for long-term follow-up and monitor their progress or lack thereof toward a disease.

We may also be able to identify family members who are at increased risk. Dr. Marjorie Shaw, a well known geneticist and lawyer, suggests that we should have legislation that makes testing in families mandatory. In her thinking, genes do not belong to individuals. We get genes from our parents and we pass genes on to our children. Therefore, they are not the property of any one individual, but they are shared with members of your family. Therefore, family members should have access to the information about your genes because of their genetic connection to you. That was the rationale used in Washington, D.C. to justify mandatory screening for sickle cell disease. This condition was classified as a communicable disease, believe it or not—communicable from parents to children. Based on that and the previous justification for mandatory screening for syphilis and other infectious diseases, the City Council instituted the law for mandatory screening for sickle cell disease in school children.

On the negative side, many who have the particular genetic marker do not get the disease even if one finds there is an 80 percent risk. That means one out of five persons in the high-risk group still will not get the disease. With regard to breast cancer, some of the things women are doing to avoid it include having bilateral prophylactic mastectomies or perhaps going on life-long chemotherapy. Some people are taking other actions to avoid getting a disease that they are supposed to be at risk for—such as having a total colectomy when at risk for the multiple polyposis gene.

Those who will not get the disease in spite of being in a high-risk group may suffer needless emotional pain. The feelings of guilt that people have can be very profound and may cause clinical depression and other kinds of behavioral and emotional problems. That can be almost as debilitating as the disease itself relative to one's ability to function in society. An individual identified with the marker may be stigmatized and questions of employability and insurability may be raised.

Population screening will be expensive; it may not be cost effective when carried out on the total population. Therefore, we should focus on the people at high risk. From a cost effective standpoint, we could screen for susceptibility. Then we can focus not on the whole population but on the highest risk individuals and therefore save money.

Family members who do not even have the markers may be stigmatized if it is known that the condition is determined by a gene that behaves as a dominant—that it is passed down the pedigree. You may be asked if there is anybody in your family with a series of genetically-determined diseases. You are required to answer that question on your insurance form and, if it is found that you gave a false response, then your insurance coverage can be declared null and void at any time. So, insurance companies do not have to test you to assess your risk; they only need to know that somebody in your family has the gene.

Ignorance of that risk could protect you from losing your insurance at some point or having your premiums raised. But again, I leave details of that discussion to others who will consider genetic discrimination.

A committee, of which I was a member, prepared a book, *Assessing Genetic Risks,* on the social and legal issues in genetic testing. After much discussion, a debate and the testimony of ethicists, these are the four items that the committee felt deserved special attention.

First is autonomy. That autonomy, a focus on the individual, should always end up being at the top of the list of ethical principles is disputed by some. We realize that there are many benefits which ought to accrue to larger groups—families, communities and so forth. There is an approach to ethics called communitarianism where the emphasis is not on the individual but on the community. In many cultures the more important value is not what happens to an individual but what happens to one's community, family, extended family, and the people with whom one lives. So some people are challenging autonomy as the primary ethical value.

Privacy is an important issue. Again there is a problem because privacy is linked to autonomy. Confidentiality and privacy are tied to each other and, ultimately, equity and fairness in treatment. The values have not changed much in recent years, but the emphasis may have changed. In this case, voluntariness is critical. We felt that, given the potential for causing harm with predictive testing, voluntariness should be paramount, and mandatory testing should be avoided for anything except for treating disease which, if untreated, would cause serious problems. So voluntariness is primary and connected to that is informed consent. The big issue here is: how do you get informed consent? What constitutes true informed consent? Whole monographs have been written on informed consent, and there is still a lot of debate about how one goes about getting true informed consent. Every Internal Review Board wrestles with that problem over and over again.

All forms of genetic information should be considered confidential. One of the reasons for this has to do with the problem of paternity or misattributed paternity. There is no such thing as nonpaternity! Every child has a father. Misattributed paternity is initially reported to the mother and not volunteered to her partner. That raises the question about the rights of the father. Of course, the only parent we know about for certain is the mother unless we do genetic testing. We counsel the mother because we know that she is a parent. The rates for nonpaternity in middle class families is estimated to begin at 15 percent at a minimum. That means the real rate is probably a lot higher.

I just want to mention DNA data banking because of the potential for harm that may come from the banking of DNA and the fact that over the long haul new problems may come up if patients are told they have other abnormal genes to deal with that they did not know about or that were not foreseen.

So where are we in predictive testing? Where might this go? I have several *Frank & Ernest* cartoons by Bob Thawes that pick up on this theme. You heard mentioned earlier about the gene for obesity. In one cartoon, Frank says to Ernest: "What's the big deal! My jeans have been linked to obesity for years!" Genes and obesity. This linkage has been known for a long time. The question is whether knowing the location of the gene that is supposedly linked to obesity will in fact do what people hope it will do: result in a way to eliminate obesity. But then there may be some benefit of genes associated with obesity. There are so many people who are obese; maybe some environmental factors selected for those genes. If we start interfering with it, we may cause other problems that were not there before. So maybe obesity is the lesser of two evils.

There are other characteristics that people are finding genes for, for example, genes for excitement or thrill-seeking. Another Thawes cartoon has Frank and Ernest in a Genetics Analysis Lab. Frank says, "Eureka! I have found the gene that causes clumsiness in people and...oops!...it just fell off the slide." And there is a gene that geneticist David Cummings claims may be associated with compulsive gambling. I mean we may soon have a gene designated for all behavior. As Dr. Botstein said: it does not matter. We'll all have a "bad" gene. We will begin to rank the genes; some genes will be more acceptable than others. Herblock had a wonderful cartoon during the height of the civil rights movements in which a dermatologist was talking to a black patient saying, "I'm sorry, sir. You have an incurable skin disease." Then of course, people are trying to blame the lack of success in society on one's genetic background. They call it IQ, but still there is a problem in using a gene as a means for explaining all sorts of things. Some researchers explain the rate of low birthweight babies in blacks on genetic differences. Why should we worry about intensive care nurseries or prenatal care to deal with those problems. They're genetic.

Of course, this extends to your families too; it does not just stop with individuals. Some fear that we will end up with committees that will sit to determine whose genes are acceptable and whose are not. I am making a huge leap here but, remember, human beings are not restricted by lack of knowledge. In fact, where we lack knowledge, we are sometimes much more dogmatic.

I attended a conference in Spain not too long ago where we talked about violence, and the bio-behavior and the social means of violence. We were presented with two syndromes associated with impulsive violence which people believe have a genetic basis. One is a serotonin deficit which is responsible and the other a monoamineoxidase deficiency that is thought to be responsible. There is a single Dutch family which has been found to have a syndrome of mental retardation, impulsive violence, and so forth. Based on this single family, people are talking about the violence gene. It does not take much to get us on a bandwagon.

Let me close with this cartoon which some people see as the ultimate in genetics—Dr. McGee talked about it yesterday—enhancement genetics. This is my favorite cartoon, *Calvin and Hobbes*. "When I grow up I'm gonna be a scientist, and I'll dedicate my career to the proposition that man can reshape the universe according to his own whims," says Calvin going downhill on his sled.

"I'll go into genetic engineering and create new life forms." His friend, Hobbes replies, "Do you wanna play God?" Calvin says, "Not exactly. God never bothered to patent his stuff."

## References

Beauchamp, T. L., and J. F. Childress. 1989. Professional-patient relationships. In Principles of Biomedical Ethics, 3rd edition. New York: Oxford University Press.

Bowman, J. E. 1972. Mass screening programs for sickle cell hemoglobins: A sickle cell crisis. JAMA 222: 1650.

Committee for the Study of Inborn Errors of Metabolism. 1975. Genetic Screening: Programs, Principles and Research. Washington, D.C.: National Academy of Sciences.

Institute of Medicine, Division of Health Sciences Policy, Committee on Assessing Genetic Risks. 1994. Genetic testing and assessment. In Assessing Genetic Risks. Washington, D.C.: National Academy Press, p. 59.

Jones, S. R., A. Binder, and E. M. Donowho, Jr. 1970. Sudden death in sickle cell trait. New Eng. J. Med. 282:323.

Lappé, M. et al., 1972. Ethical and social issues in screening for genetic disease. New Eng. J. Med. 286:1129.

Murray, R. F., Jr. 1973. The S-hemoglobinopathies: An Evaluation of Their Status in the Armed Forces. National Academy of Science, National Research Council.

President's Commission for the Study of Ethical Problems in Medicine and Biomedical and Behavioral Research. 1983. Screening and Counseling for Genetic Conditions, The Ethical, Social, and Legal Implications of Genetic Screening, Counseling, and Education Programs.

# ETHICS AND THE FUTURE OF GENETIC TESTING

## Glenn McGee, Ph.D.

Renée Fox pointed out ten years ago that bioethics has been "asocial." One consequence of this argument, for her, is that in the generation of scholarship about genetics and other medical issues, bioethicists have ignored important cultural issues in favor of philosophical examination of "in principle" problems with emergent technologies. When they had no data about ethnicity, parenthood, genetic testing, and gene patenting, bioethicists could be excused for this oversight. After all, philosophy prepares us to think at some distance about the construction of arguments, and to speculate about the desirability of different ways in which those arguments might play out in the world. But there is data now. And it is alarming because it shows that bioethics has not taken enough interest in the real-life consequences of genetic modification and testing. The future promises new uses of genetic technologies, and rather than speculate "in principle" about the just uses of new discovery, I want to take this somewhat important occasion to attempt to confront genetic enhancement "in context." That is, the issues I will discuss are among those we will face in the next few years with the commercialization of genetic tests for mutations that are related to human capacities, and with the use of genetic engineering techniques to create drugs and delivery vehicles that will enhance human capacities for traits like hair growth and ability to think with speed and concentration.

There are three reasons why genetic enhancement seems to me an inevitable part of our future. I have reviewed these elsewhere (McGee 1997, 1997b, 1997c, 1997d), and several groups (including the Recombinant DNA Advisory Commission and the Hastings Center Group on Technologies for the Enhancement of Humans) have appointed blue-ribbon panels to think about enhancement. The first reason genetic enhancement is coming is what has come to be

The author is Senior Faculty in the Center for Bioethics at the University of Pennsylvania, 3401 Market St., Suite 320, Philadelphia PA 19104. The author acknowledges assistance from fellow members of the Hastings Center Task Force on Technologies for the Enhancement of Humans.

thought of as a "proactive" approach to medicine. In the short term, with all of our current gene therapy trials in either phase one or two and none of them working especially well, proactive approaches to disease seem right now to come down to the use of genetic diagnoses for decision making prior to abortion, or the proactive use of genetic testing of pre-implantation embryos. What does it mean to prevent a disease, and how far can the public or the family go in its "proactive" approach to genetic disease and other traits? While it seems clear that the ability to use genetic tests in this way is upon us, we have devoted scant attention to the context within which genetic testing is practiced or to the moral limits of genetic testing.

The second development is more general, that of the clinical and nosological norm inside and outside genetics. Both psychology and psychiatry have seen remarkable advances in narrowing the area and scope of normalcy. Research in diagnostic procedures for isolating schizophrenia has the concomitant effect of saying something about what it means to be *not* schizophrenic, about what personal identity means for those who are normal. Much of non-genetic research has produced ideas about what it means to be normal, and these ideas of norms develop also in relation to some specifically genetic information. The most obvious examples of these in recent history are supposed genetic predispositions for aggression and male homosexuality, and tendencies to things like addiction (e.g., genetic susceptibility to nicotine addiction) and personality types. The existence of such research suggests to a public that already believes genetics is a magic bullet that we will quite soon be able to say some very significant things about genetic normalcy. That means that the potential for real enhancement exists, where things that once were considered normal become an area for molecular pathology and potentially even genetic testing and gene therapy or genetic pharmacology. And, at the same time, the "normal" has become a subject of debate in health care reform and rationing in literature, genetics, and health care economics circles.

The third development has to do with the general public. The ELSI program and others have sponsored some public education initiatives in genetics. There are now more than a dozen large scale film projects and documentaries, and these scarcely get 20 minutes into production before someone starts talking about a "perfect baby." It is impossible to avoid it. The public expects that in any conversation about the future of genetic testing, we will quickly confront what it would mean to improve offspring and what our tragic failures with such attempts in the past will mean for us in the future. The concept can also be found in scholarly literature. Bishop and Waldholz point out that genetic tests will, without question, strengthen the desire and expectation for perfect babies among affluent white middle class couples. Nobel laureate Sir John Eccles has promised, along with Leroy Hood, that we are on the road to some discovery of polygenic sites for intelligence. All of these point to the possibility of moving forward in terms of genetic enhancement. And that scares the media and many theological constituencies and results in large scale public relations disasters about perfectly reasonable advances in genetics and genetic technology, á la Dolly, the Cloned Sheep.

Clearly genetic enhancement is something that should trouble us. But the public's rationale for opposing genetic enhancement has got us off to the wrong start. I would like to review the research that has been done on families that suggests that we can do a better job of talking about genetic enhancement.

Originally, this paper was entitled "The Seven Deadly Sins of Genetic Enhancement," but my continuing experience with small groups of families and my reading of the literature on parenthood has taught me that there are really two or three pretty deadly sins, a couple of important rejoinders, and some areas where we might even want to encourage the development of appropriate kinds of genetic research and enhancement. I will describe these and talk about what they mean.

First, though, what is enhancement? Is a vitamin enhancement or is it something else? Is private school an enhancement? Orthodonture? More to the point, regardless of whether or not something is enhancement, why do we value some parts of developing the possibilities for our offspring and selectively disvalue other parts of that activity?

I think society has gotten into a morass with the definition and use of enhancement technologies because we have two *ontologies of becoming*. Both of these are fancy ways of describing meta-mistakes we make in our public discussion of enhancement. These have in no way been remedied by the ELSI program or any other public conversation about genetics because we have just not spent time thinking about them. The first one is an idea called "natural kinds and innate capacities" that allows us to hold on to the notion that certain sorts of activities are appropriate for reaching our goals, like education, and other sorts of activities are flagrantly inappropriate even if they help us achieve the same goal, for example, the use of Ritalin therapy to improve scores on tests, or the use of Prozac to improve creativity.

Why is Ritalin an enhancement if private school is not? The answer in the public's mind is that private school does not modify what Richard Lewontin calls "the bucket of human capacity." The idea of "buckets" is that we should be suspicious of any modification of our natural kind, any attempt to enlarge our nature- or God-given bucket of potential. You are well within your rights to fill the bucket, with education or vitamins, but it is something else, something "non-natural," to change the bucket's dimensions. It is on this ground that women are discouraged from cosmetic surgery, and parents are discouraged from giving growth hormones to children. But it seems obvious, at least to me, that there is a basic problem with this description of natural kinds. That is, on the genetic side, what looks like a stable genetic pattern is actually a complex matrix of interactions. What you eat contains mutagens that will change the genes in large clusters of cells. So-called "random" mutations can dramatically change the "genetic" characteristics of organisms, effecting large-scale phenotypic modifications. When you walk in the sun, your skin is affected. Radiation chemicals in the water affect germline cells and even the air you breathe right now is chock full of ingredients that will change parts of the supposedly stable blue-print of the genetic pattern. The idea that we are born with a stable bucket for any of our complex traits (like intelligence) just flies in the face of the complex and interpenetrative world in which our genes, our cells, our bodies and our minds are always changing, and in minor as well as major ways. One of the original purposes of the precursor to the Human Genome Project, remember, was to get a genetic baseline so that damage—and genetic changes—caused by nuclear weapons could be understood at a genetic level.

The more important piece of the problem with "buckets," and other natural kinds arguments in genetics, is a social problem. It turns out that intelligence, for example, gets its meaning not from some kind of acontextual set of skills that society happens to value. Lewontin points out that recent IQ tests include questions like "Who was Wilkins Macawber? What is the meaning of sudiferous?" And it is still the case that there are versions of the intelligence test administered to our children in elementary schools that include questions like: "What should a boy do if a girl hits him?" Most of what seems to be clearly clinical or objective in medicine—the definition of efficacious therapy, the idea of futility—actually turns on social values.

The problem with the ideology of natural kinds is that it begins with this notion that Ritalin and Prozac stretch a "pre-given" or "genetically-caused" bucket, whereas education only fills it. The meaning of intelligence turns out to be functional: it is a shorthand way of expressing the value of one or another kind of skill to our society at a given time. So, in fact, education and all of the other attempts to enhance our children's capacity to think are in fact enhancement goals and the only thing meaningful about intelligence or innate capacity is the question of its relation to our broader social ideas. Sometimes parents do excessive things. No question. But even the normal desires that parents hold are enhancement desires. What distinguishes the appropriate parent from the inappropriate parent is not the desire to fulfill potential versus to change a child, it is the difference between degrees of enhancement. This is why I think we find it so difficult to enunciate the reason why it bothers us when parents try to shove a particular kind of college education down a child's throat, but it does not phase us a bit when they elect to use orthodonture or insist that their children attend a particular private school. The dissonance we feel is rooted in our faith—and our doubts about the faith—that there is a clear and categorical distinction to be made between enhancement and non-enhancement.

The second ontology of becoming that is in our public literature and that gets us in trouble is the idea that, while some kinds of things we do are therapeutic, others are not appropriately medical. The idea is that we can make a distinction in our ELSI research, in our genome center research, and in nosologies between the things that will cure and the things that will improve. The idea is pretty simple, namely, that it is the business of medicine to heal people, but not to improve them.

But if this view would keep physicians from testing pre-implantation em-bryos for the *Xq28* gay gene, it would also keep them from engaging in wellness care, circumcision or infertility treatments. Plainly, we as a society value preventive medicine in some measure as crucial in the development of the physician-patient relationship—an extended conversation that involves the care of patients from the time prior to their experience of disease and other deficits, all the way to the time when they die.

The way we got to this point is sort of interesting. Part of it is the development of norms, but another is the functional similarity between investigating a disease gene and investigating a gene for an "enhancement." The laboratory that worked on $Xq28$ did not look any different than the labs that worked on *BRCA1*. The activities were similar and there is a kind of functional correspondence that suggests that, if it looks like lab research, it must in fact be medical research. This same misanalogy applies to the physician. If a test is available, if the possibility for linkage analysis is out there, if a clinician is asked a question, it is brought into the realm of medicine.

And of course, there is the social need for magic bullets. Every year, major dailies and news weeklies find ways to play up genetics as though the real promise of genetics is for enhancement and a radically different genetic future through gene therapy. Not a whole lot of attention is given to the fact that gene therapies do not work so far or that there are enormous problems, for example, involving the genetic testing that already takes place.

The issues do not really begin in the lab. Before we ask whether or not it is good medicine to offer a test for $Xq28$ or even for duplicate copies of *ApoE4* for Alzheimer's in the clinical setting, we should ask at the outset—in a new restructured ELSI program that includes a discussion of ethics in geneticists' study sections—why there needs to be the development of a particular test or a search for particular sequences in the first place. The issue that really plagues us is medicalization, a culture's aims to use clinical means to solve social problems not immediately amenable to medical solutions. In my book *The Perfect Baby*, I refer to this "pragmatic" hedge against inappropriate enhancement. An enhancement is inappropriate in a medical setting if the problem could be more directly dealt with in other ways.

Depression comes to mind. At Penn where I teach, there have been several studies of students that suggest that up to 30% are on antidepressant drugs at any given point in the semester, and this is remarkable. Certainly, depression is a real problem on college campuses. But can we treat this problem, or its progenitors, in some other way? In looking at the experience of college students in the America of today, could depression be partly due to the fact that these same students, after their 300 person classes, go home to watch their "Friends" on television and to eat frozen dinners? Even if there is a depression gene, the search for it seems to me to be a kind of genetic equivalent to the StairMaster. We hope that we will be able to continue on our current social and institutional tracks, gorging ourselves in the daily faith that we will burn it off by climbing therapeutic stairs to nowhere.

## A Pragmatic Approach to Enhancement

The idea I want to offer is that parenthood gives us the wisdom to provide answers—including answers to questions about genetic enhancement. We do not need an exotic genethics. We do not need a bioethics of individualism to look for the answer to genetic enhancement. We have only to turn to the lessons of parenthood. Those lessons are transmitted in the public "self-help" literature.

I want to make three points. Parenthood is really the social institution in our culture where enhancement is debated. It is also the first frontier of enhancement technologies. The first wave of this proactive genetic medicine that Leroy Hood talks about will be the decisions parents make about genetic tests in prenatal contexts, for example, the *BRCA1* test. Finally, dangers of enhancement are not unique to enhancement because enhancement is not unique to genetics.

Parenthood teaches us that there are three deadly sins which may be considered dangerous. The first one is *calculativeness*. It is dangerous when parents emphasize plans for their children. After all, they might not work anyway. Think back to your own childhood. What really taught you the lessons that you remember as a part of the development of your character? I still remember a little bit of the carefully calculated speech my father gave me when I was 13 or 14 about what it means to be good and wise in the context of a community that would make broad demands on me. However, when I think about social justice and what it means in relation to my family, I really draw on an argument my parents once had when I was in the back of the car. That conversation shaped my character in unexpected ways. Divorce does the same thing and, of course, the death of a mother or grandmother. We cannot anticipate the ways in which our character will be shaped by weird happenstance.

Calculation in parenthood, when it works, can be an oppressive yoke to bear. Whether it is piano lessons forced upon children until the age of 20 years, or just pictures from a teen magazine, our parents and culture transmit ideas that take their toll on growing children. The danger of genetic enhancement technologies in this regard is that, whether or not they come to work, they create a profound expectation.

Moreover, the celebration of a child's sharing with the family can be undermined by what may be referred to as trait-based planning, and the calculativeness that comes with systematic enhancements could and frequently will be much too narrow. The parent who gives the growth hormone ends up placing too little emphasis on other kinds of growth in the child. Then there is the key point. Children do have some right to grow up under the burden of their own ambitions and moral mistakes. This is not to say in a radical and individualistic sense that children have a right to their own natural kind.

It is a fact that children grow up in families, that families appropriately produce the moral atmosphere in which children grow. It is a subtle lesson, but it is one that we can all relate to from our own experiences. If you calculate too much, it will not work and, when it does work, it might work too well. The thing that a child remembers about his parents' calculation may be not only that he did not want to be a football player or a doctor but that the calculations were actually in part a creation of a matrix that ensured failure.

The second deadly sin is *shortsightedness*. Who would have thought that in the last decade an empire would be destroyed, that the United States would go to war with a Third World country, that genetic testing would be seriously developed, or that the stock market would crash three times, each time foretelling an impossible decimation in the economic system? Who would have guessed that George Bush would lose? Well, it is the same sort of problem that plagues eugenics. In 1925, Mueller gave a lecture in which he described a set of people that we ought to begin breeding ourselves towards. They included Lenin and Marx. But in Texas he subsequently gave a lecture in which he included not Lenin and Marx but Descartes and Lincoln. The point is that hasty judgment is a real problem when you are trying to describe archetypes for enhancement.

Kids are incredibly malleable; they can change and adapt. Just look at how quickly they moved out of business school when it became apparent that environmental studies was the best way to get a job. Even with parents pushing them, they found a way out of the Masters of Business Administration. However, if you change the design, you lose the malleability. When the genetic optimists begin talking about the long-term future of the gene, and *Time* and *Newsweek* say that we will have children one day with gills and scales, I wonder whether or not the ocean would be such a great place to live. You would not want to live in New Jersey's bay area with gills and scales. When it is suggested that my child might benefit from intelligence gene research, similarly I wonder what today's intelligence will be worth tomorrow. Look at what happened to Einstein's physics, and yet most parents report that they would chose male offspring if presented with the choice and many—more than 40% in the last study—indicated willingness to improve intelligence through genetic modification.

In the final analysis, the real danger is that of *hasty judgment* per se—the last of the deadly sins. I grew up in Waco, Texas, which is among other things about 30 miles away from Texas A & M, the land of genetic enhancement of plants. Former Texas Agriculture Secretary Jim Hightower states in his book *Hard Times, Hard Tomatoes* that what has been discovered in the genetic enhancement of fruits and vegetables is that what works in the lab does not usually work in the field. There are no "controlled" trials of fruit enhancement. Moreover, compared to the original varieties, the more flavorful and more interesting varieties of genetically enhanced plants and animals almost always become much more vulnerable in the field to pests.

In terms of humans, how difficult will it be to design something that will really work, how difficult to live an engineered life. Even if one could live such a life, even if the present attempts in our culture to use Ritalin, Prozac, private schools and vouchers to engineer our children are successful in some measure, the question for those children is: What will they think of the imperfect world in which they live? Will they scoff? Will they be sensitive? Will they have social and political vulnerabilities? What will they understand?

In the final analysis, it seems that there are several not so deadly sins here. Our task in examining the possibilities of genetic enhancement is not to figure out whether or not informed consent can be obtained from those who are in gene therapy products for conditions that are not yet identified as diseases or for those who would avail themselves of the opportunity for an *Xq28* genetic test. It is not, anyway, *primarily* about informed consent. The primary issue is not whether or not enhancement will bring obvious harm to our culture. Rather, the task is to determine whether or not some strategies can be developed in thoughtful ways, in terms of our family and our cultural ideals, about what it means to improve our children. We all want to improve our children, as part of the responsibility of being parents and members of society. But how? Without good schools, the child who has an intelligence gene will just be a child who is great at selling drugs on the street.

Society's goals and thinking about genetics have been too narrowly defined. Too much of the Human Genome Project has been left outside of the scope of normal peer review for molecular genetics and completely outside the scope of review for social and political issues. There must be a new combination of thinking about society and thinking about genetics. Scientists must be trained to do this, and responsibility must be taken for what is said to reporters and how we guide the public conversation about these social issues.

## References

Fox, Renée. 1978. Experiment Perilous. Oxford University Press.

Hightower, Jim. 1989. Hard Times, Hard Tomatoes. New York: Penguin.

Lewontin, Richard. 1993. Biology as Ideology: The Doctrine of DNA, Oxford University Press.

McGee, Glenn. 1997. The Perfect Baby: A Pragmatic Approach to Genetics. New York, Rowman & Littlefield.

McGee, Glenn. 1997b. Parenting in an era of genetics. Hastings Center Report, March/April.

McGee, Glenn. 1997c. Obstetrics, genetics, and the future of the family. OrGyn, June.

McGee, Glenn. 1997d. Subject to payment. JAMA. July 16.

# GENETIC DISCRIMINATION IN HEALTH INSURANCE AND WHY IT IS WRONG

## Thomas Murray

I want to begin with two texts: One of them is inscribed on a coffee mug that my children gave me for Father's Day a few years ago. The mug reads, "Insanity is inherited; you get it from your kids." Parents laugh, but parents—particularly those who have adolescents—understand what is true about that statement. Even if the genetics are completely wrong, there is nonetheless something existentially very powerful about the concept. But that is not what I am going to focus on. I am going to focus on another text you have heard before, but I will bring it up to date at least in one sense. The statement is: "All persons are created equal." The original form was: "All men are created equal." All the people writing it were men and, in fact, they did not quite mean it, because they left out large categories of male persons. But the thought is one of our primary American dogmas. It is not a scientific claim though. It is not a claim that all people are biologically indistinguishable. It is a claim rather, that, in a political community, we must treat all persons as if they have guarantees of moral, political and legal equality. That is what this statement means to me.

To put that in context, one way of seeing human genetics is perhaps as the leading example of the science of human inequality, that is, the science of human differences. If roughly one in 200 women carry mutations of the *BRCA1* gene—which puts you at about an 85 risk of breast cancer, that is a biological difference. Human genetics shows us subtle and not so subtle ways in which we are different from one another. Why is that important? It is important because the challenge facing us as a community is how to live up to our commitment to moral, legal and political equality in the face of human difference.

Aristotle wrote that distributive justice consists of treating like cases alike and different cases differently. The problem is, of course, figuring out which likenesses and which differences are relevant. What I want you to notice is that,

The author is a professor of Biomedical Ethics and director of the Center of Biomedical Ethics at Case Western Reserve University (10900 Euclid Ave., Cleveland, OH 44106-4976) and a Fellow of the Hastings Center, a pioneering research institute in the field of bioethics.

159

for certain purposes such as health insurance, genetic differences have become candidates for "filling in the blank" of what makes people different. So back to a woman with mutated forms of *BRCA1*. Is she different for the purpose of deciding whether or not she should have access to health care?

Along with Jon Beckwith and Robert F. Murray, Jr., I was one of the founding members of the National Institutes of Health's ELSI (Ethical, Legal, Social Implications) Group working with the Human Genome Project to think about ways in which genetic information might have important social roles. I want to add a footnote here: I would not have been a part of ELSI if I did not think that genetics and genetic information can do people a great deal of good. In fact, one of the main reasons for having the Task Force on Genetic Information and Insurance was because of the concern that people—either because of good reason or misinformation—were afraid to get genetic information that might be useful to them and their families, or were refusing to get it because of the loss of access to insurance. The potential benefit of the work in human genetics would not be achieved because people would be afraid that they might be left out of something that is very important to them, health insurance—something they might in fact have more need of because they have a genetic risk of disease.

It certainly looked as if one of the areas in which genetic information would become relevant to people would be insurance. So the ELSI Working Group decided to set up the task force, which first met in May, 1991. On the task force were people who represented groups of individuals who were at risk of or had genetic diseases, people representing the insurance industry, and a sprinkling of experts. Our job was to come up with a report in two years. There were many issues, some of them technical. Was genetic testing a likely possibility? Some issues had to do with how insurance works. What are the practices of insurance companies? We learned what we could about these questions and relied on the people who represented these different areas to bring in their own background and knowledge.

Some of the issues were moral ones. Those were not necessarily more difficult than the technical and scientific ones. They were the crucial problems we had to tackle. Just to put this into perspective, we needed to know how hard it was, technically, to get genetic information about people. As you may know, it just takes a little scraping of the cells from inside the cheek to get enough cells to generate a DNA sample. I got a new insight on this at a genetics meeting a couple of years ago, from the then-director of the FBI laboratories. He mentioned that his laboratory had completed the work on the samples from the World Trade Center bombing. They were given the envelope and letter that were sent claiming credit for the bombing and were able to link the cells on the back of the stamp to one of the suspects. He said they also learned that the person who licked the envelope was a different person. As you might guess, one of the largest collections of DNA samples in the world belongs to Publishers Clearinghouse—although I don't think they realize it. Just think about it. You lick the stamp and all those little stickers for extra bonus prizes and then even put your name and address on it, what more could they want! I don't mean to frighten you but getting DNA samples is not a big technical problem.

Shortly after the task force began its work, we became aware of a split among members over different understandings of what it would mean to have a just health care system. Back to Aristotle. What would it mean to have distributive justice in health care? What should decide who gets access to health care, on what terms, and at what price? I confess that I came at this from a different point of view than that of the representatives of the insurance industry. But I think it is very important to understand their way of seeing this. There is a concept called "actuarial fairness." The *Harvard Law Review* (1986) had a very good piece by Karen Clifford and Russell Iuculano who explain and defend the concept of actuarial fairness. Actuaries get data about the likelihood of particular outcomes based on certain variables. If I am a male, I am less likely to live as long as those who are female. That is an actuarial statistic. Women live longer than men. Should we be allowed to use that in deciding how much to charge for health insurance and how to figure out pensions? That is a social question. But actuarial fairness is a moral claim. It is a claim that it is morally just to have people pay according to their likely cost.

Let's try this approach. Go back a few years. Pretend you are the owner of a few oil tankers. When your fleet of oil tankers is steaming in and out of the Persian Gulf at the very height of the Gulf War, you are very happy that you are getting paid a lot of money for this service. Now pretend your friend is made the owner of a second fleet of oil tankers. This second fleet happens to be cruising the safest route in the world, wherever that is. I am Lloyds of London, but I am solvent and both of you want me to write insurance for your ships and their cargoes. I say fine, and I will charge you both exactly the same rates. Now, you are in and out of the Persian Gulf; how do you feel about that? No problem, you will take that deal. But your friend? She is in the safest route in the world. How do you feel about that? She thinks she is overcharged. Is she? I think she is. It makes reasonable sense, good moral sense in fact, to subscribe to actuarial fairness in such cases. Actuarial fairness becomes a good description of justice in certain cases like that one.

The insurers who were part of our task force did a good job of educating us about the industry's conception of fairness. They used slogans like: You do not insure a house that is on fire. They said, if you were willing to insure houses that were on fire, then people would not come to you until they start smelling smoke. Why pay premiums until you have to?

Using the concept of justice as actuarial fairness, someone who has a genetic disease is a "house on fire." Someone who has a genetic risk of disease is a "house that is smoldering, but the flames are not yet visible." I want to leave you with those metaphors in thinking about actuarial fairness.

Now I am going to take you down another path. It is an exercise about the different notions of the nature of justice. We just heard one: actuarial fairness. Let me now give you a different one. You are going to have to help me on this. Let's say I represent the Cleveland Browns. Many people know that the Cleveland Browns, or should I say the team formerly known as the Cleveland Browns, is now in Baltimore and is called the Ravens. There still is a team called the Cleveland Browns. It is now 1999, and the new Cleveland Browns have begun to play. You have to bear with me; remember, this is all hypothetical. But I am talking to you because I represent the Browns and you are the committee that decides who is going to receive the Super Bowl Trophy for 1999. I want to persuade you to give the Super Bowl championship trophy to my franchise.

You may not have been aware of it, but this 1999 version of the Cleveland Browns is the most virtuous football team in the history of professional sports. These 45 athletes, even when they are not practicing or playing, are very dedicated. When they are not home doing the laundry, the ironing, making meals, watching the kids—when they are not doing things that otherwise occupy their time, they spend the rest of their time visiting children's hospitals. Some of them take an occasional course on Shakespeare. This is the most virtuous group of football players ever assembled! I ask you: will you give the Cleveland Browns the Super Bowl Championship? Who votes to give them the trophy? What? No one?

Okay, let me try another tactic. You know there are teams in cities like Miami where, if they don't win the Super Bowl Championship in January, they will go and sit on a sunny beach and enjoy the weather. No problem. But, in Cleveland where the city is united with the team more emotionally than in any other city I've ever known, this championship is needed desperately. As I said, if Miami doesn't get the Super Bowl championship, big deal, they'll go for a swim. If Dallas doesn't get it, big deal, they've won too many already. But Cleveland! Cleveland needs it more than any other city! January is not our best month, you may have heard that. It gets pretty cold. Let's just say that I could persuade you that the Cleveland Browns really did need the championship more than any other city. Who would vote to give it to them? What? No one again? This is a tough group.

I have one more approach. I will pay you more money than any other team representative if you give Cleveland the Super Bowl championship. What about it? I see a couple of votes out there! This approach uses the George Steinbrenner theory of sports championships. If George could buy it, then he probably would, but the rules do not permit that so he just tries to buy stars to win them.

What is it going to take to get the Super Bowl championship? I was afraid of that; it is going to be a long Cleveland winter, I think, in 1999. We have a new franchise, and I have a feeling we are not going to win the Super Bowl any time soon.

You are right, obviously. We understand how it is that the Super Bowl Trophy gets distributed. It goes to the winner, just like other things distributed on the basis of merit. If we found that the Pulitzer prizes were given not to the writers who wrote the best books but to the writers who had the cutest wink or the biggest checkbook with which they were bribing judges, we would think that there had been a corruption. We would consider it horribly unjust to give prizes on the basis of anything other than merit. That makes sense. That is how we distribute that kind of social good.

There are other goods. Let's say I go to an electronics superstore and say, "I'd like that VCR over there, and I think you should give it to me because I'm really a very nice person. I am a virtuous person. Would you give it to me?" They would show me the door. Would that be an injustice? If I ask a salesman, "What is it going to take to get the VCR," he is going to tell me to give him the money for it.

Is that unjust? No, we feel that many kinds of goods should be distributed according to what you can afford to buy. Once you have the money and you realize it is what you want, then you can have it. It is not an injustice to say to people who cannot afford it for one reason or another, or do not have the money for a VCR, that they cannot have it.

What about health care? How do we understand the just distribution of health care? Let me give you a dramatic example. Say that I have just learned about the availability of a transplantable kidney. How should I figure out who gets that kidney? We actually have a system in place. The United Network for Organ Sharing helps make those decisions, and they have certain rules to follow. Who should get it? The one who needs it the most? The one who is going to pay the most? Suppose I were a very eccentric, wealthy person, and I thought the best fertilizer for roses was human kidneys, and I was willing to pay more than anybody else for that kidney. Should I be allowed to buy it? No. We don't really think that in the end the distribution of health care, particularly life-saving dramatic health care that makes a difference in people's lives, ought to be by the market. We think that health care ought to be distributed according to need.

Food also ought to be distributed according to need. It turns out that most people get enough food because we can produce it relatively cheaply and distribute it widely. When people cannot get enough food on the market because they cannot afford it or because it is not available, we see that as famine or distress. We try to help them out. So need lies underneath the market when it comes to the distribution of something like food.

Health care might work somewhat similarly. We might say that so long as market solutions like private insurance work in getting health care to the people who need it in a reasonably efficient and fair manner, then we will accept that. We will also step in and take care of the people who are not served by that system, because in the end, need is the underlying principle. We are willing to be fairly open about what set of social systems we use to satisfy that need. But in the end, the underlying moral imperative is *need*. That is my argument.

That is not the argument of actuarial fairness. In fact, it is the reverse of it. Actuarial fairness says that the more acute your need, or the more expensive your need, the less willing we will be to provide it for you, and certainly not at the same price as someone else whose anticipated need is less than yours. The apotheosis of actuarial fairness in health care is: we screen you—not next year but sometime in the future—to pick up a series of different genetic risks for, let's say, cancers and perhaps other diseases. Then we will tell you after the screen that we have good news for you, you probably have no special genetic risks for A, B, C, or D cancers, but you do have a risk for breast cancer. If you have *BRCA1*, it is not certain that you will get breast cancer. There is a probabilistic connection, however, and that is what actuarial science is about—probability. So we will write you an insurance policy, and it will insure you if you are run over by a truck or have a heart attack, or get one of those other cancers but, if you get breast cancer, tough. That's the moral logic of actuarial fairness.

But insurers, certainly the ones in our group, were not and are not heartless people. They agreed that need was the underlying moral dimension. What we disagreed on was, and it was not exactly a disagreement, the role of genetic information in private health insurance. Since we were a task force and not charged with the redesign of the world's health care system or the U.S. health care system, only to address genetics and access to health care, we focused just on that piece of the puzzle.

I need to tell you one more story. The thing that we wrestled with early on, was whether genetics was different from other health care risks. There is talk of "genetic exceptionalism," that is, treating genetics as an exception. But the question really is: "Is genetics different?" We looked hard at that and, in the end, we rejected it. I think we rejected it because it subscribes to what I have come to call "the two-bucket theory." In that theory, one bucket has a "genetic" label on it, and the other bucket says, "Non-genetic." Now give me a disease or risk factor and tell me which bucket to place it in. If the disease is Huntington's disease, which bucket? Genetic. Getting hit by a truck? Non-genetic. We expect it will always be non-genetic. Cholesterol level? Genetic...and non-genetic. Okay. We have a problem here. If you took what people said or what you read or what television advertisements said about keeping your cholesterol levels low, the answers would include: diet, stay away from saturated fats, don't eat chocolate cake. It is really painful, but you do a lot of things to keep your weight down. You also exercise. These are indeed ways you can modify your cholesterol level but, if you look at what proportion of the variability of cholesterol level is due to lifestyle variables and what percentage is due to genetics, which do you think is the higher percentage? It is genetics. Or so geneticists consistently tell me.

I'll give you my sister-in-law's example. A lovely person. She can eat a carton of Häagen-Dazs ice cream and not get her cholesterol level to budge above 150. If my wife or I gaze longingly at one, we are in the mid-200s. It has to do with genetics. But there is a problem. It is also true that you can in fact lower your cholesterol level or raise it by what you choose to do. So which bucket do I put it in? It doesn't fit well in either. The "two-bucket theory" does not work for most of the diseases that are important in terms of how many people's lives are affected. Most significant diseases and risk factors are going to be like cholesterol—both genetic and non-genetic. Where does that leave us? We could pretend that genetics is different. Or we could accept the fact that it is probably not that different, that genetics ought to be seen as part of a continuum of factors that might predict later illness. We could then say that genetics is a subset of the kinds of health risk factors that insurers might be interested in. The task force's conclusions can be summarized under two points: (1) there should be no individual underwriting in health care coverage, i.e., all the health risks from genetics and otherwise should not be figured into whether you have access to insurance, what you are covered for, and how much you will be covered for; and (2) genetics should not be used for deciding who gets access to health care or health insurance. Genetics should not be treated differently from other health risks. Some will object, aren't there certain things you can do to hurt yourself? The answer is yes. Are people morally responsible to take care of themselves? Yes. As an ethicist, I would be existentially inconsistent to say you have no moral responsibility for your own welfare. You do. But this is a policy question, isn't it? How should we deal with that knowledge? I think one can give different answers and have them all be reasonable, but an answer that is not reasonable is: we underwrite you for a whole range of things but deny you access to the health care and insurance you are most likely to need because of a genetic or other risk.

Needless to say, there was some controversy within the task force. We ended up with not quite a unanimous vote in favor of the report. The national Blue Cross/Blue Shield organization signed the report; the Health Insurance Association of America neither signed nor opposed the report. I think quite reasonably they saw there would still be a role for a good private insurance industry even in a world of no underwriting. There are other ways to compete. With large groups there is no underwriting anyway. If the world is moving toward the insurance of just large groups, then we would not have to worry about underwriting.

The only group that opposed it was the American Council of Life Insurance. I felt a little sorry for their representative on the task force. He had been a part of our conversations for two years and of course had helped write the report. When his board looked at the language about underwriting, I suspect they were not happy.

I want to finish by describing some of the response in terms of legislation on the state and federal levels. As I said, the Task Force on Genetic Information and Insurance issued its report in 1993. At that time we thought there might be national health care reform. We all know what happened to that. If in fact health care reform had passed and done away with individual underwriting, most of our concerns about genetic discrimination in health insurance would have disappeared. But it did not happen. Instead, what we see are at least two trends. One has been a series of state laws directed toward genetic discrimination in health insurance.

State laws are for the most part very well-intentioned. I have talked to people who have been drafting these laws so I can speak firsthand about their intentions. But I have to say that, for the most part, the legislation has been feckless; that is, the laws don't do much. Why? Two reasons. The first is they adopt a narrow definition of genetic information—simply, the results of a direct DNA test. Right now that amounts to something very tiny, but it is something. In fact, even looking at an enzyme downstream from the supposed gene of interest would not be considered a direct DNA test. If you check for anything other than the gene's DNA sequence or even if you check the things it makes, it is no longer a direct DNA test. Most of the state laws focus on the results of the direct DNA test.

The second reason can be illustrated by my own state of Ohio. Ohio was considering a law that would permit insurance companies to offer products based on an individual's bringing in the results of his/her DNA test. This is how it would work. If sometime in the future you want to go out and undergo a cancer screen, insurers could offer a product to you that gives you a discount much like being a non-smoker gives you a discount. What I think might happen is the insurance industry will see an opportunity here. Normal competition will take place among companies. They will begin offering products for people who have lower risks in actuarially significant ways. What will happen to companies that do not offer such products? High risk people will flock to those companies, and the companies are either going to have to go out of business or raise their rates. Or else the companies will join in offering products for people with low risk. If genetic information proves actuarially significant and if insurers are permitted to use the information when people bring it voluntarily, you will see a segmentation into low risk and relatively high risk underwriting. That is one reason to think that the only real, stable and just solution is to do away with individual underwriting.

There is also something called the "Genetic Privacy Act," drafted, I believe with ELSI program support, by George Annas and his colleagues at Boston University as a proposal for national legislation. The GPA is also being adopted by at least some states, and a version of it has been introduced in Congress. But this too uses a very narrow definition of genetic information—again, the direct DNA test results. The authors realized they had a problem. They had to either take on a definition which would be broad enough to really capture all of what we regard as genetic discrimination, but that would seem to run into what we are calling the "two-bucket" problem. So they opted instead for a very narrow definition on the grounds that at least that could be contained. George Annas knows my views on this. I respect their effort, but I wish they had not taken that route. There is a saying in ethics that "the best is the enemy of the good." That is, if you have your eye on a far-off goal, i.e., the elimination of all genetic discrimination in health insurance, you might lose sight of the small steps you can make toward it. I think that Annas and his colleagues can claim, with good reason, that they are trying to make those steps toward it. So, I respect what they are trying to do, but I do not think that, in the end, it will solve the problem of eliminating genetic discrimination in health insurance.

I want to close with a warning and an option. First, the warning. I happen to be very excited about the Human Genome Project and the kinds of scientific advances and the good effects it can have on human health. I am an enthusiastic supporter of the Project. I will be honest about that. Do I think there are some ethical issues that need to be attended to? I certainly do. One of the things that

I am worried about is not any specific issue, not even the use of genetic information by insurance personnel. I am most worried about a kind of overselling of genetics, especially using genetics to explain complex behavioral and social phenomena—and most especially when it is used to "apologize,"that is, explain away, enduring social inequalities. We have been there before. We had an overselling of behaviorism for a while. We will have overselling of other ideas.

Right now, genetics is really exciting and for good reason, and the press picks it up and writes about it and pretty soon the public becomes convinced that in fact there will be a gene for everything and that genes will explain a great deal. Well, genes will explain a considerable amount, but they won't explain everything. They are only one of many factors that make us what we are. I have been saying this for many years, that one of our saving graces has been that those things which are humanly most important to us—courage, creativity, loyalty, and the like—are the things we are least likely to have genetic explanations for.

Persons who have fought so hard for genetic privacy have tended to emphasize and, I think, overrate genetics as special and as threatening. George Annas came up with the very evocative metaphor of "a future diary." That is how he sees the human person's genome: a future diary where your fate, in significant measure, is written out for you. I looked hard at this notion of a future diary in a chapter in a book edited by Mark Rothstein, and I lay out what I think are some of the problems with it. And I offer an alternative. In George's version, the diary is already written.Your diary contains your intimate thoughts, which is one of the reasons why we want to keep strangers from peering into it. We use diaries to describe our most private, significant, and intimate thoughts and experiences.

We could instead think about an individual's genomic inheritance as being the physical volume on which the diary is inscribed. You can buy different diaries. Some of them have relatively few pages, and some of them have many pages. In the same way, some of us will have relatively shorter lives, and some of us will have longer lives. In diaries, the quality of the paper differs, and it will be harder to write on some of them, just as there are going to be the tough periods in our lives because of some problem that our genes may contribute to. Let's regard our genes not as a diary, but as a list of the obstacles—and opportunities—we are likely to encounter or perhaps as a somewhat improved prediction of how long we will have to do what matters to us—to be with the people we love and to accomplish the tasks we have set for ourselves.

Our genes no more dictate what is significant about our lives than the covers and pages of a blank diary dictate the content of what is written within. Our genes might be regarded metaphorically as the physical but blank volume on which we will write our diary. The content, though, we write ourselves.

## References

Clifford, Karen A., and Russell P. Iuculano. 1987. AIDS and insurance: The rationale for AIDS-related testing. Harvard Law Review 100:1806-24.

Murray, Thomas H., Mark A. Rothstein, and Robert F. Murray, Jr., The Human Genome Project and the Future of Health Care. 1996. Indiana University Press, Bloomington, IN.

Roche, Patricia, Leonard H. Glantz, George J. Annas. 1997. This volume, pp. 187-196.

Rothstein, Mark A. (Ed.). 1997. Genetic Secrets: Protecting Privacy and Confidentiality in the Genetic Era. Yale University Press.

Task Force on Genetic Information and Insurance. 1993. Genetic information and health insurance, National Institutes of Health, National Center for Human Genome Research, Bethesda, MD.

# INSURANCE CONCERNS: ARE THE FEARS EXAGGERATED?

## David J. Christianson

My topic today is genetic testing and insurance. I chair the American Academy of Actuaries Task Force on Genetic Testing and Life Insurance. While I work for a fraternal benefit society selling life and health insurance, I am not representing my company nor am I representing the insurance industry today. My intention is to deal with this topic from the standpoint of a professional actuary, drawing on the work of our task force, the Academy's published Risk Classification Statement of Principles and the Actuarial Standards Practice. My area of expertise is in individual voluntary insurance, especially life and disability insurance. Thus, I will limit my remarks to those areas. Joan Herman will later cover individually purchased medical insurance.

I would first like to tell you a little bit about the American Academy of Actuaries. Its mission is "to ensure that the American public recognizes and benefits from (1) the independent expertise of the actuarial profession in the formulation of public policy, and (2) the adherence of actuaries to high professional standards in discharging their responsibilities."

The strategic directions of the Academy are stated as follows:
- Identify, assess, and prioritize key public policy issues;
- Identify, establish, and strengthen Academy access to key public policy-makers;
- Provide guidance to the profession's research areas so as to develop research projects supportive of public policy initiatives; and
- Provide the technical expertise of the actuarial profession to decision makers on key public policy issues.

At the conference and prior to it in newspapers and journals, fears have been raised about the potential for discrimination in insurance from use of genetic information. Some of the fears raised are:
- If insurers find out I have a genetic condition, will they:
  - cancel my coverage?

A Fellow of the Society of Actuaries, the author is vice president of the Insurance Services Division for the Lutheran Brotherhood, 625 Fourth Ave. So., Minneapolis, MN 55415. He currently chairs the American Academy of Actuaries Task Force on Genetic Testing and Life Insurance. © May not be reproduced in any form without the expressed written consent of the author.

• raise my premiums?
• deny my application for future coverage?
• Due to insurance fears, will people not seek out genetic tests?
• Will insurers invade my privacy?
• Insurers just want favorable risks. Will they exclude many from coverage?
• Will I be forced by insurers to take genetic tests and find out information that I do not wish to know?

The purpose of my talk today is to examine these fears and, while acknowledging them, assert that these fears are often exaggerated, that alternatives exist to lessen the impact and that working together to find solutions will produce the greatest public good.

To begin, it will be important to discuss voluntary insurance and the role played by risk classification. In crude terms, you can think of insurance as a bet being placed between an insured and an insurer. The actuary is the oddsmaker and the underwriter assesses the information about the individual seeking insurance, placing the individual in the appropriate risk class.

To help you understand this, let's work through a few scenarios. Then I will come back to tie it all together. How many of you play the lottery? Why do you do it? I imagine it is because you feel the money you pay is reasonably related to the risk, and you have the *same chance* to win as everyone else. Let's examine this more closely. Suppose we were to have a 100 person lottery. Each person puts in $10, therefore winnings equal $1000 (assuming there are no expenses).

1. Each person gets a ticket. I put the stubs in a hat and draw the winner. Seems OK. Most would play.

2. Now let's say I predraw the winner and post it. You get to see the ticket number before you buy. You choose to buy or not. I assume only one person would want to play. The $10 collected would not cover the $1000 paid out. Where the chance is certain, $1000 must be charged.

3. Let's keep the rules the same as number 2, but I include the ticket cost in the registration fee for this meeting—no choice. Obviously you all are playing, and it will work out financially. It may not be fun. You may grumble like Social Security participants often do, but it works.

4. Finally, let each person who chooses to play get one ticket for $10. However, if you live in Minnesota, you get 20 extra tickets. Also, if you are _____ (*you fill in the blank!*), you get 20 extra tickets. Suddenly, some people have a greater chance to win than others. Would you play? You get 20 times more chances of winning, so you should pay almost 20 times more, or about $200. Note, too, that since there are more tickets eligible to be drawn, everyone else's chances to win decreased.

Now let's make it more specific to life insurance. Suppose two of you were to bet on how long a person would live. Take an older person. If you were each to place $100, would you bet the person would live more or less than 20 years? How about a 35-year-old woman? What more would you need to know? Age, health status, previous history, family history, and more. What if the bet were between a 40-year-old man and you? Would you feel any differently? When we started, each bettor had equal knowledge. Who has the "knowledge edge?" What if the other person knew that they had lung cancer? Would that affect the bet?

Insurance companies, self-insurers and governments, anyone that runs and administers insurance programs, face these issues. They collect money from participants, pay expenses, pay out benefits and perhaps make profits. All of this must come from the money collected. It must balance. To get people to participate, the costs must be reasonably related to the risk or, alternatively, people must be forced to participate as in lottery example #3, where the cost was added to the registration fee.

Risk classification is a fundamental precept of any sound private, voluntary insurance system. It becomes less important in public, involuntary insurance systems. The purpose of risk classification is to group similar risks having reasonably similar expectations of loss. Think back to the lottery example. You would be most likely to "play" if you had the same chance of "winning" as everyone else. If the odds are tilted away from you, you are less likely to participate. If you are forced to play no matter what, the system can still work. However, it may be highly inequitable. That is because the players with no chance of winning are still forced to contribute in order to "subsidize" the predetermined winner. So, while the system *works*, it does not produce a *fair* result.

According to Actuarial Standards of Practice, a sound risk classification system should be based on four principles: First, risk classification should reflect cost and experience differences based on relevant risk characteristics. Extra risk translates to extra cost. Financially, it does not matter *why* these differences exist. It just matters *that* differences exist. To work properly and to be perceived as fair, the costs and the resulting premiums must be in balance.

Second, the system should be applied objectively and consistently. By this principle, for example, all males of the same age with similar health characteristics should be charged similar rates for life insurance.

Third, the system should be practical, cost-effective, and responsive to change. This means there are limits on how much effort and money can be spent to classify a given risk. And, risk classification systems are dynamic; for example, when polio was eliminated as a public health hazard, the system changed to reflect that development. Today, coronary conditions and high blood pressure are being underwritten far more liberally than 15 years ago. There are many other examples. Personally, I have an inherited genetic condition—von Willebrand's Syndrome, a clotting disorder. Twenty years ago, I was offered life insurance at 100% to 200% extra mortality charges. Today, due to more knowledge of this condition by the insurance companies, I am a standard risk.

Fourth, and perhaps the most crucial, antiselection should be minimized. Antiselection is an actuarial term that requires some further explanation. Applicants for insurance often know more about their own risk factors than the insurer can learn in the application and underwriting process. However, a sound risk classification system should limit the ability of the applicant to have an unfair financial advantage at the expense of the insurance company and/or its insureds. This unfair advantage, in essence, is what we mean by antiselection. An example of antiselection might help. Suppose a person has an anonymous test and finds out they are HIV positive. They become concerned that they will not be able to keep working so they purchase disability insurance. They do not reveal the information to the insurance company, so the contract is issued "standard." Their odds of collecting benefits are great. Their premiums will be low. This is not a level playing field. It is antiselection.

Although each of these principles has implications for genetic testing, two deserve special mention: The system must be practical, cost-effective and responsive to change. Insurers will need to be sensitive to costs of genetic testing. It may be too expensive to use many of the tests. So there is some built-in unlikelihood of global screening by insurers. Other tests may be widespread in usage and inexpensive. As medical practices and tests change, underwriting practices will need to keep up, adding tests when cost-effective and useful, especially when not using the tests will promote antiselection.

This leads to the second point. Antiselection may well be the biggest concern. In a private, voluntary insurance system, where there is a choice to buy or not buy, and especially where the amount and time of purchase can be selected, the rating structure and indeed the overall financial soundness can be at stake.

For example, it is not unusual to find term insurance rates of $2 to $3 per $1000 of insurance at age 35. You do not have to be an actuary to know that this implies that deaths of under 2 per 1000 are expected. It is easy to see that if people know they have serious health problems and are allowed to enter the group at the same price (perhaps by withholding genetic testing information they have) that the prices will be inadequate, causing either financial failure of the insurer or big price increases. Thinking of the lottery examples, healthy people quickly figure out inequities and refuse to play. Those who would pay less than the risk assumed are attracted. A price spiral ensues. One only needs to look at health insurance problems to see real examples of this, where healthy young people drop out because they feel the premiums they pay will greatly exceed the benefits they would collect. Joan Herman will talk more about this later.

Let's now narrow this down to the implications of genetic testing. As you have seen, risk classification is the means by which insurers analyze each applicant. The goal in individual life and disability insurance is to place the applicant in a pool of risks having the same probability of death and thereby charge the same premium. Insurers look at a variety of factors such as age, sex, weight, blood pressure, family history, etc. Insurers get only one opportunity to evaluate the risk. After that the contract is in force and no changes can be made.

Two questions come to mind. Do insurers need to have applicants take genetic tests to properly evaluate risks? Do insurers need to see test results of which applicants already have knowledge? The answer to the second question is "yes." As you have seen, if the applicant has such information he or she may antiselect. The result will be felt on insurers, other policyholders, and other potential applicants, as premiums become insufficient to cover the costs. In essence, a whole new set of fears will arise, such as: Will healthy people have to pay more for insurance? Will costs increase for existing policyholders? Will insurance be available? Will government have to intercede and require or provide coverage? As to whether insurers need to have applicants take genetic tests, it is not clear that this is needed, especially if one assumes a narrow definition of genetic tests. With the current knowledge base, the system is working fine right now. It may be wise, though, not to close down this avenue. Being responsive to change can help expand coverage and reduce costs, too.

Let's now return to the fears. *"If insurers find out that I have a genetic condition, will they cancel my coverage or raise my premiums?"* This simply cannot happen in voluntary individual life or disability insurance. Once the contract is established, it is in force as long as the premiums continue to be paid. Also, the premiums are established at issue. In some cases, a guaranteed maximum premium is set and premiums are charged at a level lower than the guarantee. However, the premiums can only be changed for the group as a whole, not for individuals. Genetic testing has no bearing here.

*"If insurers find out that I have a genetic condition, will they deny my application for future coverage?"* It could happen. As mentioned earlier, 4% of all applicants are currently denied coverage for life insurance, far fewer at younger ages (age 55 and younger) and more at older ages. So some of these same people may be denied coverage based on genetic tests, too. Will a lot more people be denied if genetic results are known? I think not. Let me explain.

1. Genetic tests (absent increased suicides or depression) will not increase mortality or morbidity rates. The opposite effect is likely to occur. As people learn about their predisposition to disease, they will take steps to monitor, prevent and deal early with resulting conditions.

2. Insurers make money by insuring people. There is no motivation to decrease the pool of applicants.

3. Competitive forces are intense. It is not easy to sell life and disability insurance. There are over 1,700 individual life and health insurance companies and over 200,000 agents. There are companies that specialize in placing coverages that other companies deny. I believe the competitive forces will minimize the exclusion of coverage on people due to genetic test results.

4. Genetic tests may "clear" people who otherwise would have been rated or declined due to poor family history or inconclusive data.

5. Innovation is possible.

Let me give you three examples that bear these ideas out. In the U.S., applicants who are HIV positive are generally denied coverage. However, recently an insurer began offering coverage in South Africa for HIV positives. The premiums are approximately equal to those that a healthy 75-year-old would have to pay. That makes sense, because their mortality patterns are quite similar. This demonstrates that innovation is possible.

Second, anyone with a family history of Huntington's disease is generally denied coverage today until they reach an age at which it is clear they will not develop the disease. With a negative genetic test, such a person would be eligible for standard coverage immediately. Furthermore, with a positive test, it is possible coverage could be offered, such as an endowment or even a life insurance policy. A five-year-old with the gene could expect to live about 40 to 45 more years on average and thus perhaps insurers could offer coverage at premiums near to what a healthy 50-year-old would pay. Obviously this coverage would be far more expensive than for other five-year-olds, but coverage *would* be available.

As a third example, fears have been expressed that anyone testing positive for *BRCA1* will be denied coverage. Using advanced mathematical models (Monte Carlo method), Dr. Marty Engman and C. Allen Pinkham studied expected mortality patterns in such people. The conclusion, shared last February at the Genetic Issues Seminar: Update 1996, held in Atlanta, showed that the extra mortality resembled the increase in mortality due to poorly controlled hypertension in women or the difference between male and female mortality. Of course, the pattern of elevated mortality will not be level and hence the extra mortality will be only slight at some ages and in some situations and quite a bit greater in others. The bottom line is not to expect everyone with the *BRCA1* gene to be declined or even highly rated for life insurance.

Let's examine another fear. *"Will insurers invade my privacy?"* There are two concerns here. The first is that insurers will see one's medical records. This happens routinely today and is a necessary part of the underwriting process to avoid antiselection. This is not the real fear. What will be done with the information is the fear. I've discussed that fear already. Second, there is concern that insurers will pass along that information to others. Of course, that could happen, but insurers are heavily regulated and highly encouraged to protect information and respect privacy and confidentiality. Although this is not an actuarial matter, I believe the industry has an excellent track record in this regard. In fact, I called the Minnesota Department of Commerce, and they have registered no complaints against insurance companies in the past 10 years about not keeping confidential information private.

*"Due to insurance fears, will people not seek out genetic tests?"* This fear could be reduced or eliminated in a variety of ways. First, people could purchase insurance before they take the tests, and provided they were otherwise good risks, they would obtain coverage. Researchers could build this cost into the cost of their research and buy group coverage for all study participants. In fact, it puzzles me why this is not done because, if this fear is real, a biased sample could result if not dealt with. Alexander Tabarrok of Ball State University has suggested mandatory genetic insurance. This is insurance that would insure against a positive test result. He proposes that every individual be required to purchase it prior to being tested.

Another fear is that *"Insurers just want favorable risks. They will exclude many from coverage."* The facts do not bear this out. Only 4% of applicants are currently denied coverage for life insurance. In fact, insurers want to cover *all* risks that are willing to pay a premium commensurate with the probability of loss. In competitive industries one rarely finds companies voluntarily giving up market share.

The last fear is *"Will I be forced by insurers to take genetic tests and find out information that I don't want to know?"* As I said earlier, I do not believe there is a big current need to require testing that involves new genetic tests. This need is far less than the need to learn of test results the applicant already has received. Also, the narrowly defined genetic tests would likely be too expensive to be cost effective at the current time. However, as tests become inexpensive and widespread, the need to test will grow.

In discussing the fears, I have mentioned what is and what is not possible. However, this may seem like an incomplete, hazy answer to those facing decisions today about whether to be tested. On the other hand, the fears and *real* concerns are probably a bit hazy and out in the future, too. There is not a multitude of tests currently available for which there are long lines of people waiting to take them save for the insurance issue. There is not clear agreement either that such tests will be desired or be conclusive in predicting or preventing disease so how quickly will tests become widespread?

So what next? One approach is to ban genetic testing for life and disability insurance. It has already been banned to some extent in some states for health insurance. No state, to my knowledge, prohibits underwriting based on genetic information for individual life and disability insurance. However, New Jersey recently passed legislation that any underwriting action must be reasonably related to anticipated claims experience. This seems reasonable and is in accord with Actuarial Standards of Practice. Let's examine the expected results, though, of an outright ban on use of genetic information.

First, you must understand that any restrictions in risk classification have a cost. The cost will be borne through some combination of: higher prices, lower profits, restricted access to/availability of coverage, and shifts from a voluntary private system to an involuntary public system of coverage (in other words, mandatory participation). The cost is directly related to the degree the risk classification process is restricted. In genetic testing, for example, the degree of restriction may be related to the "tightness" of the definition of genetic tests upon which legislation is written. Similar costs occur for any restrictions to underwriting.

Even though you may feel differently about the public policy implications, it makes no difference cost-wise whether an increased risk of death or disability comes from: (1) prior accidents causing health problems; (2) self-imposed risks such as smoking or being overweight; (3) diseases you acquire; or 4) conditions you inherit. Think back to the last lottery example and you will see that it does not matter *why* someone got 20 extra chances. The result is the same—a system out of balance. To the insurer, extra chances = extra cost.

This brings up a second issue, fairness. Why should a person with a genetic predisposition to a disease or disorder be given preference over others who acquire a disease, disorder or disability due to environmental factors, accidents or the like? If some are given preference, others must pay more to make up the difference. Is that fair to them? Should we create a privileged class of people born with defective genes? Do we provide insurance at standard rates to the person with *BRCA1* but deny coverage to the woman who has breast cancer but is not a *BRCA1* carrier?

Third, how can insurers practically identify the "pure" genetic disorder? That could turn into an underwriting nightmare!

Fourth, if the intent is to disallow use of genetic information, the result may be that a person with such a condition would have no incentive to buy early or to restrict the amount of coverage purchased. A person whose health is going downhill and where disability is imminent would want to buy disability insurance. And what is to prevent large "death bed" purchases in the case of life insurance? These are like the second lottery example. A $100,000 life insurance contract bought in this fashion would require a $100,000 premium. At any other price, this would quickly ruin the system.

If genetic testing is to be disallowed, the result then may be higher priced insurance in the short run, fewer issuers, fewer offerings, and a move to guaranteed issue, restricted, high premium insurance or even government-imposed coverage.

The other approach is to let the current system continue to function and work in partnership to find creative solutions. There is certainly room to provide unique coverages within the voluntary insurance system. Government can help too by providing regulatory and individual tax relief when unique products are designed and by considering ways to provide backup coverages.

Keep in mind, too, that the presence of genetic factors may often be of no significance for added risk of death or disability. Standard issues may be available. For example, consider me with von Willebrand's. Where conditions exist that present extra risk, the same alternatives are present as for anyone else whom the underwriting process identifies as an extra risk:

1. Rated (higher-priced) coverage, where premiums reflect the added risk assumed.

2. Guaranteed issue insurance, typically with high premiums and restricted coverage for the early policy years.

3. No coverage.

4. Automatic group coverage through employers.

5. Social Security coverage.

Finally, another option for most people is to buy coverage young and to buy guaranteed insurability options. Doing this *before* testing is done, or impairments occur, provides the availability of coverage in the future.

Working in partnership, new solutions can be explored as well. They include the following:

1. Closely examining the mortality and morbidity patterns of persons with genetic traits in light of current treatment, expected monitoring and proactive health management. This analysis can bring clarity to the underwriting process.

2. When costs or expected underwriting outcomes are known, communication of this to affected populations can reduce fear of the unknown.

3. Shifting thinking in the life insurance industry from chronological-age underwriting to a biological-age model. Let me explain. Currently, it is likely that a 30-year-old with a 15-year life expectancy would be declined for life insurance. Take an HIV-positive person for example. On the other hand, a healthy 75-year-old may have the same life expectancy and be offered coverage. In chronological age thinking, the extra mortality is expressed as some percentage of the standard mortality. In the case of HIV in a 30-year-old, the percentage exceeds the maximum extra percentage that will be accepted by most companies. A shift in thinking could allow coverage to be offered, albeit expensive relative to other 30-year-olds.

4. Offering other products where the likelihood of early death is an advantage in the pricing of the product. Substandard payout annuities are rare, but actuarially make sense.

5. Allowing individuals declined for life insurance to accumulate money in a specially designed product having the same tax advantages as life insurance. This would mean no income tax on accumulation and no income tax when the proceeds are paid to the beneficiaries.

6. Considering developing a genetic insurance policy (as mentioned earlier) whereby all people to be tested pay a premium and the coverage promises insurability to those who test positive.

7. Developing government-sponsored pools to offer limited amounts of insurance to those denied coverage.

8. Offering limited scope products to people with short life expectations. For example, accident-only policies could be offered. The difficulty would be separating out accidents from suicides.

These are a sampling of the possibilities. The American Academy of Actuaries stands ready to help explore these options. I truly believe that this avenue may have the best long-term payoff to society.

In summary, I believe that the fears, though real, are exaggerated. Quick fixes, such as prohibiting access to genetic information, will lead to new fears and problems that will affect a much greater group of people. On the other hand, alternatives already exist in today's environment that help alleviate the concerns. Working together to find new creative solutions will produce the greatest public good.

# GENETIC TESTING AND INSURANCE: A QUESTION OF BALANCE

**Joan Herman**

While I work at Phoenix Home Life, I am here as a member of the American Academy of Actuaries' Genetic Testing Task Force. So I am not representing my company or the insurance industry. I am speaking as a professional actuary, and the positions I take may or may not be in accordance with the industry's position.

In talking about insurance, I will confine my remarks to voluntary insurance because that is where most of the issues related to genetic testing and insurance will arise. Insurance is voluntary if the purchaser has the option to buy or not to buy, and they may also have as their choice the amount of insurance to purchase. This is in contrast to something like Medicare or Social Security, where participation is mandatory, and the amount of coverage is set by a formula so there is no individual choice.

I am going to go through some actuarial principles—not moral principles or social judgments. Actuarial principles are really ways of describing how a private, voluntary insurance market works and what kinds of consumer behaviors arise.

Insurance is a mechanism for transferring and spreading risk. However, not every form of risk is capable of being insured. Generally speaking, there are five conditions which should be met for an event to be considered insurable. First, it must involve uncertainty with respect to occurrence, timing and/or severity. That is, you have to be uncertain as to whether the event will occur, or when it will occur and/or what the magnitude of the loss will be, if in fact loss does occur. Second, the risk should display what we call statistical regularity. Simply put, that means that if the same sequence of events were to occur over and over again, a pattern of probabilities would begin to emerge. Third, the fact that the event has occurred should be definitely determined. We should be sure it has really happened. Fourth, the occurrence of the event must cause an economic loss for someone. Finally, the fact that the event is to occur and how

The author is a member of the American Academy of Actuaries Task Force on Genetic Testing. She works at Phoenix Home Life Insurance Co., P.O. Box 5056, Hartford, CN 06102-5056.

severe it can be should not be something that can be known or controlled by the person being insured.

When all these conditions are present, an insurance mechanism can be used to spread risk. This is done by pooling risks faced by a large number of persons with a similar expectation of loss, paying benefits to those who experience the insurable event, and funding those benefits by receiving premiums from the entire group of insured. One of the key phrases in what I just said is the pooling of persons with similar expectations of loss. I get asked an awful lot of times why insurers have to underwrite and classify risks. They usually put it this way: " I thought insurance was about pooling of risks so why can't you put everybody into one big pool, charge them all the same rate, and not do any underwriting?" Of course, what is missing is that crucial phrase—similar expectations of loss. Remember, we are talking about the pooling of persons with similar expectations of loss.

The question is: why does that matter? It has to do with what is called adverse-selection or anti-selection. The terms are synonymous. They refer to the tendency of people who have a greater than average expectation of loss to apply for insurance protection to a greater extent than other people. If you know you are at a higher risk of premature death, you will tend to buy more insurance than someone else who believes himself or herself to be healthy. For example, if you just found out that you had cancer and you could go out and buy insurance at the same price as when you did not know you had cancer—at the same price as other healthy people, chances are you would go out and buy it. And if at the point you got the diagnosis you did not have health insurance, chances are that one of the first things you will do is run out and get it.

Some states require insurers that write individual health insurance to issue coverage to anyone who wants to buy it at standard rates with little or no limitation on pre-existing conditions. In such states, we see adverse selection occurring from conditions that range from AIDS to cancer to pregnancy. There was a recent article in the *Wall Street Journal* reporting that a large health insurer in a state with such a law received a letter from a satisfied customer who obtained coverage under the provisions of the law. The woman praised the insurer for providing benefits to her and her newborn baby, and added that now that she had the baby she was canceling coverage. She went on to note that she would surely come back to the insurer for coverage if she gets pregnant again. From the woman's perspective, she is simply doing what makes economic sense. But from the insurer's perspective, this was adverse selection.

Insurance is supposed to be about an individual's willingness to pay a reasonable amount in order to protect against a possible but uncertain economic loss. But in the example just described, we have a situation guaranteed to produce an economic gain for the insured: she knows she is pregnant, she knows the approximate health care costs associated with the pregnancy, and she knows the cost of the premiums. She can sit there and calculate that, if she buys the coverage for just that little bit of time, she comes out ahead.

From the actuary's point of view, that is a real problem because, when we design a product, our first and foremost obligation is to make sure that premiums are sufficient to cover the promised benefit payments, so that we can pay out those benefits to everybody we made a promise to. In the example I just gave you, there may be no way to do that. Many will say, "Well, raise the premiums." If we raise the premiums higher so that the person no longer gets an automatic gain, premiums will be so high that most healthy people will drop their coverage. When they drop out, we do not have enough money to cover the benefits of the people left. If we raise the premium again and again, we end up with a sort of vicious cycle called an "assessment spiral," and then things tend to just fall apart.

It is important to know that the consequence of adverse selection is not simply reduction in insurer profits. That can be its initial impact—and sometimes its only impact. But if you cannot bring it under control, then it will lead to insurer losses and eventually the insurer's insolvency and its inability to meet its benefit obligations. So then adverse selection has the potential to totally undermine an insurance program.

How do we minimize adverse selection? Insurers try to do that by grouping people into relatively homogenous categories with respect to the likelihood and expected magnitude of loss. That is, we group together risks with similar expectations of loss, and that is what we call risk classification. When we do that, the premiums for each risk class can be set at a level commensurate with the groups' expected cost, and the people in the class should feel, we hope, that they are paying a reasonable amount for the protection they are receiving.

The main tool you have in classifying these risks and thereby minimizing adverse selection is information. If the insurer knows as much about the risk as the applicant does, he should be able to appropriately classify it. So insurers gather relevant information before issuing a contract, and that process of gathering information and using it to classify risks is what we call underwriting.

Underwriting will vary depending on what kind of coverage is being requested and how it is being paid for. When you underwrite a home-owners policy, you want to know what the house is made of and how far it is from a fire hydrant. For health insurance, you are usually getting the person's health information.

How the insurance is being paid for also affects the underwriting process. There are two main categories particularly in respect to health insurance: individual insurance and group insurance. In individual insurance, the premium is typically paid for by the insured who receive benefits. In group insurance, the employer usually pays a significant part of the cost.

With individual insurance, the underwriting process generally involves gathering information about the individual to be insured. Group insurance, on the other hand, relies more on the information about the group as a whole rather than on specific information about each group member. The insurers may want to know the ages of the members and may be concerned about the group's total claims over the last several years in terms of dollars paid out. But in most cases, they are not going to ask anything about the health history of individual members in the group.

One may wonder why it is that, when underwriting for individual health insurance, information about the individual's health is needed, but it is not needed for underwriting group insurance. The answer to that revolves around the amount of choice the individual has, because the more choice the individual has, the greater the opportunity for adverse selection.

In individual insurance the purchaser has total choice whether to buy at all and, if so, what to buy and how much to buy. For example, you can buy a $10,000 or $1,000,000 life insurance policy or nothing. In addition, the purchaser bears the full cost of the insurance. So if an individual purchases a family health insurance policy that costs $7,000 per year, that is all coming out of the purchaser's pocket. Now let us contrast that with group insurance.

With group insurance the employer is usually making the decision about which coverages are offered, and the decision is usually based on business considerations such as what sort of benefit package the employer feels he or she has to offer in order to attract a high quality work force. If the employer pays the full premium, then all the employees are automatically covered and there is no individual choice. If the employees have to pay some of the premium, then the choice is generally limited to whether to participate in the plan or not. They usually cannot choose the amount of benefit and generally they have to decide in a limited amount of time whether to join the plan.

In group insurance, typically employees are restricted with respect to the timing of the purchase and the amount of coverage. And they are not paying the full cost of coverage. All these features tend to reduce the risk of adverse selection, and this allows the insurer to rely less on information about individuals and more on information about the group as a whole. So the general rule is that, the more the individual controls when, if and what to buy, the greater the risk of adverse selection and therefore the more information the insurer needs in trying to make sure that the premiums are commensurate with the risk.

Returning to the issue of genetic testing and my topic of "Genetic Testing and Insurance: A Question of Balance," in dealing with the interplay of genetics and insurance we are trying to balance a variety of legitimate concerns. There are concerns about privacy and confidentiality, concerns about the availability and affordability of health insurance and other kinds of insurance, as well as concerns about adverse selection, financial solvency and viability in voluntary insurance programs. So how do we balance these concerns.

Some would argue that the best solution is to simply prohibit insurers from using the results of genetic tests as a part of the underwriting process. When I say prohibit them from using the results of tests, I mean more than just saying an insurer cannot require you to take the test. What is also being recommended by some people is that an insurer should not be able to inquire about the results of a test an applicant previously had taken.

If we prohibit that, then we have dealt with the privacy and confidentiality concerns but we would not have addressed the concerns about adverse selection and financial solvency. While this point may not be obvious, it may also not do a very good job of addressing the issues of affordability and availability. At first glance, prohibiting the use of genetic test results may seem to guarantee the availability of affordable insurance to anyone who has inherited a genetic predisposition to disease. However, in reality, if you legislate prohibitions to risk classification that results in increased adverse selection. you end up with a combination of higher prices and fewer choices in the short run. Depending on how severe the adverse selection becomes, you can end up with virtually no choice and prices so high that virtually no one can afford the coverage. This really does happen.

For example, some states have tried—with the best of intentions—to improve the availability and affordability of health insurance by requiring insurers who write individual health insurance to accept all applicants regardless of health status or age at a single community rate. They thought that, since it works on the group insurance side, then individual health insurers should be able to treat all their individual insurance as one big pool and not need to underwrite or charge differential rates. But as I discussed earlier, there are important protections against adverse selection that operate in the group insurance context that simply do not exist in the individual health insurance market.

Many of the states that passed such laws are now seeing that they are not working as hoped. They are seeing rates climbing year after year as healthier lives drop out of the system. They are seeing fewer choices of plan designs available on the market. For example, in Kentucky where reforms went into effect in July of 1995, we understand that 50 carriers have left the state and only one insurer is left writing individual health insurance throughout the state. This carrier is experiencing large losses and seeking a rate increase of 27.5%.

Now a law that prohibits insurers writing individual health insurance from charging differential rates based on health status or age is not the same as a law that prohibits an insurer from using the results of a genetic test. You could ask how much impact would such a prohibition have. First, there would be virtually no impact on group insurance because group underwriting does not typically involve asking for individual health information. On the individual side, the impact of such a prohibition will depend on the exact definitions of genetic information and genetic testing that are used. How broad or narrow are those definitions? If they are defined very broadly so that we are in the kind of situation where the definitions would preclude an insurer from asking about family history or the results of a blood cholesterol test, then all medical underwriting could be virtually wiped out. If you eliminate all medical underwriting in the individual health insurance market, the results will be severe adverse selection problems in that marketplace.

If you take a much more narrow definition, and probably, if you were to ask me right now how big of an impact there would be, I would say: not really big. We probably would not have the kind of an upheaval where the individual market starts to collapse. Why? Because insurers are not asking for that kind of test and, more importantly, they are not only not asking explicitly for the results of these kinds of tests, but also there are not many people in the population who have had these tests. This is why this is not an issue.

The question becomes: what happens several years down the line? At that point there may be a lot of people who have had some kind of genetic test and, if we prohibit an insurer from asking about the results of the test that someone already had, you can get an imbalance of information, where the applicant knows a whole lot more about their health than the insurers are allowed to know. That can create an adverse selection problem. That is the kind of situation where the issue could become important. We do not know that is going to happen, maybe it will not, but that is the risk. Given this, it is worthwhile to look at the issue and whether there are other options to consider.

When examining the overall situation, it is important to remember that, with respect to health insurance, for those under age 65 the overwhelming majority of people in this country get their insurance through their employers. Assuming they enroll when they first become eligible, they are not subject to any individual underwriting. Less than 10% of the health insurance in this country takes the form of individual health insurance. Furthermore, the passage of the Kassebaum-Kennedy legislation ensures that, if people lose their jobs or lose coverage as a spouse because of divorce or something else, or switch employers but have been covered in the past under a group insurance plan, then they are guaranteed that they can get new health insurance provided they did not have a break in coverage for more than 63 days. Therefore, one of the major ways that in the past people found themselves both uninsured and uninsurable has been dealt with.

What about the person who has never been covered under a group insurance plan? In over 40 states such a person would be able to get health insurance regardless of health status from one or more of the following: a high risk pool, an individual health insurer, or a Blue Cross open enrollment plan, with high risk pools being the most common source of guaranteed coverage.

High risk pools do typically charge insureds more than a standard rate but a lot less than their true expected costs. The difference is financed by a subsidy mechanism. Most typically, this is an assessment done on all health insurers, but it can also be accomplished through more broad-based financing such as a payroll tax or a hospital tax. There are additional options, which are not in place yet, that may improve universal access within a private, voluntary market. These options have been outlined by the American Academy of Actuaries monograph called "Providing Universal Access in a Voluntary Market," and copies may be obtained from the Academy's office.

Other options include a universal system where everyone is entitled to have coverage from cradle to grave. In terms of Dr. Thomas Murray's remarks that it should be provided based on need, this can only be guaranteed in the ideal sense through a universal system. This could be accomplished through a government plan like Social Security, or through the private sector with participation being mandatory. The recent debate about health care reform in this country has shown that there is no consensus yet to implement this type of reform in the United States, but it is obviously one of our options.

It is noteworthy that options such as high risk pools or universal coverage systems have the advantage of treating all persons with medical problems more evenhandedly. One of the things I find troubling about prohibiting the use of genetic testing in insurance underwriting is that it gives preferential treatment to just certain high-risk people. Is it fair that someone who is a quadriplegic because he got hit by a drunk driver or that someone who is HIV positive might be unable to get insurance at standard rates, while someone who has the gene for Huntington disease would be able to go out in their late thirties and buy lots of insurance at standard rates? I really do not find that that fits in with my sense of fairness, so I would like to find alternatives that do not preferentially discriminate in favor of just certain high risk people at the expense of others.

As we learn more and grapple with the issues, I think we will find there are additional options. First, in many cases a genetic condition may not result in increased risk. In other cases it may be that, if someone modifies his/her behavior or takes some preventive measures, there may be no increased risk and coverage could be offered at standard rates.

In still other cases we may find that, if a person applies for insurance at a young age, the extra mortality or morbidity risk may be small and spread out over many years, and so could be issued for just a small extra fee. I do not know for a fact that anyone is doing this, but from what I know generally—without having done all the actuarial calculations, it seems that someone with a *BRCA1* gene who tries to buy individual health insurance when she is 18 or 20 years old would probably be able to get this insurance at close to a standard rate.

I hope that by sharing our knowledge and our respective expertise and by putting our heads together to look for creative solutions, we can create workable options that address all the various concerns. The American Academy of Actuaries stands ready to participate in the process.

# THE GENETIC PRIVACY ACT:
## A PROPOSAL FOR NATIONAL LEGISLATION

**Patricia Roche, Leonard H. Glantz, George J. Annas**

Privacy is a major issue in medical law, and genetics is a major force in contemporary medical science. Nonetheless, the combination of these two fields has only recently been seen as central to both individual rights and medical progress. Disclosures in June of 1996 that White House officials had wrongly acquired and read FBI files of raw background checks of prominent Republicans reminded Americans that there is no such thing as a completely secure and secret file of personal information. While not specifically discussed, the addition of a DNA profile or simply a DNA sample to such a file could give the person with access to it much more additional information about the unsuspecting individual, information that could be used against the individual without his or her knowledge. In late June, 1996, Senator Pete Domenici (R-N.M.) introduced S.1898, the "Genetic Confidentiality and Nondiscrimination Act of 1996" (GCNA) which was based in large part on the "Genetic Privacy Act of 1995" (GPA) drafted by the authors.[1] The purpose of this article is to outline the purpose and provisions of the GPA. Along the way we will also highlight some of the differences between the GPA and the GCNA.

## The Human Genome Project

The Human Genome Project (HGP) is an international effort to create a high resolution picture of the human genome, and to sort genes, the informative sections of DNA, from the vast regions of DNA which appear to have no function. Molecular geneticists theorize that decoding the information in this human blueprint will ultimately advance our understanding of every function of the human organism that is even partially affected by genetic inheritance.

Patricia (Winnie) Roche, J.D., is Assistant Professor of Health Law, Leonard H. Glantz, J.D., is Professor of Health Law, and George J. Annas, J.D., is Professor and Chair of the Health Law Department, Boston University Schools of Medicine and Public Health, 715 Albany St., Boston, MA 02118.
 This article first appeared in *Jurimetrics*: Journal of Law, Science and Technology 37: 1-22 and is reprinted by permission. Copyright © American Bar Association. All rights reserved.

The genes responsible for several diseases including Huntington disease and cystic fibrosis have already been located. Decoding of other sections of the genome has identified changes in genetic material which seem to increase susceptibilities toward other diseases, such as some forms of cancer and heart disease. While the manifestation of those diseases in individuals is likely the result of undetermined interaction between genetic and environmental factors, the ability to read sections of the individual's genome contributes to knowledge of the risks that an asymptomatic individual has of developing disease in the future. If the speculation of some geneticists proves correct, the human genome may also tell us about the development of a wide range of behavioral characteristics such as alcoholism, aggressiveness, risk taking - and even happiness.[2]

The tasks of decoding the entire genome and sorting scientific genetic facts from science fiction will probably take another decade, but that has not inhibited the desire to read the bits of the genome that have been decoded to date. Testing kits for a mutation of a normal gene called *BRCA1*, which is associated with increased risk for breast cancer, are already being marketed regardless of concerns about the misunderstanding and misuse of information derived from test results. What the presence of discernible mutations in such genes means in terms of disease prediction is not known for certain. Nonetheless, the analysis of a person's DNA will have a significant impact on how individuals view themselves and are viewed by others, and this raises serious issues of control of genetic analysis and genetic information. Who should decide about what genetic tests, if any, should be performed? Who should have control of DNA samples which contain the individual's entire genetic blueprint? Should all genetic information be treated the same way? How should access to private genetic information be controlled? These are some of the questions presented by this new capability.

Task forces and committees convened by ELSI (Ethical, Legal and Social Issues working group of the HGP), various professional associations and state legislatures have focused on the legal and ethical issues presented by the Human Genome Project. Throughout their discussions it has been generally recognized that present laws will not adequately protect the information that will be created in this new genetic era. Exactly what form or scope new laws should take, however, continues to be debated.

The initial question is whether genetic information is so unique that it should be treated differently from other types of medical information. If genetic information was not significantly different from other highly personal information, particularly medical information, we could simply amend current laws governing the confidentiality of other sensitive information so as to specifically include genetic information. The primary feature of DNA that argues for legislating additional safeguards is that the information contained in a DNA sample can be used to predict the likelihood that an asymptomatic individual may suffer from a variety of conditions in the distant future. In this way the individual's genome can be thought of as a coded probabilistic future diary.[3] Knowledge of such predictive information may be possible only through decoding the individual's DNA, but once created, this information can profoundly affect the individual's self-perception and life choices. Educational pursuits, career decisions, even retirement planning can be based in part on genetic factors. Reproductive choices may be based not only on the individual's risk of suffering from a major illness, but the probability that a particular genetic disease or disease susceptibility will be passed on to the individual's offspring. Decoding an individual's DNA also divulges information about that person's parents, siblings, and children, and can therefore affect how family members perceive and relate to one another as well.

The greatest risk to individuals as the result of the creation and disclosure of genetic information, however, comes from the use that governments and others who control resources can make of such information. As history has shown, genetic information and misinformation has been the basis for vicious discrimination against those viewed as undesirable or unfit because of genetic status. Because all genomic information about an individual is contained in each DNA sample extracted from that individual, stored samples will provide more genetic information than was imagined when samples were originally taken (as our knowledge of the function of various genes increases). Threats to individual privacy and autonomy are therefore presented not only by the performance of currently available genetic tests, but by the longtime storage of samples that could be subject to additional analysis in the future. Thus DNA storage issues must be considered separately from issues regarding the storage and release of information derived from DNA testing.

As a result of these concerns, in 1993 we undertook the development of guidelines for DNA banking which would protect the privacy of individuals whose DNA was stored under contract with the DOE portion of ELSI. As we worked on this issue, however, we concluded that DNA banking was a small part of the genetic privacy puzzle. We concluded that it was necessary to regulate the acquisition of DNA and its analysis, as well as the storage of collected DNA samples. Consequently, and with input from and final endorsement of the ELSI Working Group, we crafted a much broader legislative proposal, the GPA. Though written as a proposal for federal legislation, it can also serve as a model for state legislation, and as a source of rules and regulations that could be voluntarily adopted by anyone engaged in the collection, storage and analysis of identifiable DNA samples.

The overarching premise of the Act and its comprehensive set of rules is that no stranger should have or control identifiable DNA samples or private genetic information about an individual, unless that individual has specifically authorized the collection of DNA samples for the purpose of genetic analysis, has authorized the creation of that private information, and has access to and control over the dissemination of that information.

## Core provisions of the GPA

The core provisions of the Act are that: no collection of DNA for analysis is permissible without an informed and voluntary authorization by the individual or his or her legal representative; those conducting DNA analysis are prohibited from doing so unless execution of written authorization by the individual or legal representative has been verified; no analysis exceed the scope of the written authorization; DNA is the property of the individual from whom it is obtained; DNA samples be routinely destroyed once the authorized analysis has been completed; anyone who holds private genetic information in the ordinary course of business keep such information confidential and is prohibited from disclosing it unless the disclosure has been authorized in writing by the individual or legal representative.

These rules govern all circumstances in which individually identifiable DNA samples are collected for analysis. Collection and analysis of DNA without prior authorization is permissible only for law enforcement identification activities when otherwise authorized by federal or state law (the GCNA properly adds the U.S. military to this exception)[4] and for the identification of dead bodies, provided that the kind of analysis that is conducted is limited to DNA typing.[5] Such profiling analysis results in information useful for identification but which is otherwise without meaning. It is similar to fingerprinting in that it involves unique individual markings that reveal no informative details about the individual who is the source of the fingerprint. Use of DNA typing for identification, therefore, does not result in the creation of the kind of genetic information which needs protection, and the exception for this purpose does not conflict with the underlying premise of the GPA.

Prior authorization from the individual is also not required when DNA is collected pursuant to a court order, provided that the order is drafted according to the statutory specifications.[6] This exception could be invoked in paternity actions and other cases in which the genetic condition of a party is at issue. But even in these cases, genetic analysis is limited in such a way that it would rarely, if ever, result in what we consider private genetic information. In most such instances, identity is the issue.

The prohibitions and restrictions on collection and analysis of DNA apply to activities involving *identifiable* DNA samples.[7] Consequently, enactment of these rules would not affect use of non-identifiable DNA samples for research because there are no privacy issues involved when samples are not and cannot be linked to an individual. This is perhaps the major difference with the GCNA, which is based on the premise that all DNA is "identifiable" through DNA fingerprinting and thus individual consent must be obtained for all research.

**Authorization Required**

To facilitate informed decisions regarding DNA analysis, the GPA prescribes several verbal disclosures that must be made before written authorization is obtained.[8] Requiring verbal disclosures provides persons to be tested with the opportunity to ask questions and to obtain additional information they deem necessary. These disclosures are designed to include information regarding the potential benefits and risks of proceeding with an analysis. Therefore, the individual must be informed about the information that can reasonably be expected to be derived from the analysis, and the utility, if any, that such information might have for that individual. Although the average person is familiar with having tests that measure a physical condition, such as blood cholesterol levels, it is unlikely that most people would anticipate the implications that undergoing a genetic analysis might have, not just for themselves, but for others as well. Therefore, the GPA requires that the person be informed that he or she may discover something that is also of importance to genetic relatives and consequently will have to decide whether or not to share that information with family members. We view the duty to share needed information with family members as a moral obligation, and nothing in the GPA legally obligates anyone to share his or her own private genetic information with others.

Because of the interest that others such as insurers or employers may have in the results of a DNA analysis, the person must also be forewarned of the possibility that others may condition future benefits on the disclosure of information resulting from such analysis or from the very fact that such an analysis was performed. The GCNA specifically restricts the *use* that employers and insurance companies can make of genetic information. Use restrictions were beyond the scope of the GPA, but they can provide important protection as well. Use restrictions, however, are after-the-fact and difficult to enforce (as the FBI files case shows), and cannot be a substitute for privacy protection.

In addition to receiving such verbal disclosures, before deciding whether to proceed with the collection and analysis of DNA, individuals must be provided with a notice of rights and assurances.[9] The final procedural requirement before collection and analysis of a sample can legally proceed is the execution of a written authorization, and the information that must be included for such authorization to be valid is described in detail in the Act.[10] For example, the purpose of the analysis and the identification of the entity that will perform it must be explicitly stated in the authorization. Without informational requirements such as these, the substantive rules created by the GPA could not achieve their purpose. Compliance with these requirements will create a "chain of custody" of identifiable samples so that individuals can effectively exercise the rights established by other substantive provisions. For example, the GPA provides that an identifiable DNA sample is the property of the individual who is its source, and grants the individual the right to destroy the sample at any time.[11] Enactment of these provisions would be of little use to individuals unless they were made aware, not only of the existence of the legal rights and obligations created by the GPA, but where their DNA samples are located. Therefore, a copy of the authorization must also be provided to the individual[12] and any facility storing the sample which intends to transfer its operations to someone who will use the sample for a substantially different purpose than was originally authorized, must notify the sample source and provide an opportunity to reclaim or order destruction of stored samples.[13]

## What Genetic Information Is Protected by the GPA?

The definition of "private genetic information" in the GPA is based on our conclusion that there is a subset of genetic information that is unique and should be particularly guarded. This subset, termed "private genetic information," is: any information about an identifiable individual that is derived from the presence, absence, alteration, or mutation of a gene or genes, or the presence or absence of a specific DNA marker or markers, and which has been obtained: (1) from an analysis of the individual's DNA; or (2) from an analysis of the DNA of a person to whom the individual is related.[14]

This definition excludes information about obvious and manifested genetically determined characteristics such as height, eye color, and symptoms of some diseases. This information is excluded because publicly accessible genetic information cannot be considered private. In addition, this definition excludes information about genetic conditions which would be knowable from the development of family medical history or through biochemical analyses, such as blood tests for the presence of gene products. For example, an individual's status as an asymptomatic carrier of a gene associated with a particular disease may be inferred from knowing that certain relatives suffer from that disease. Similarly, the probability that a young woman has a genetic predisposition to breast cancer may be deduced from examining the occurrences of breast cancer in her family. Both of these pieces of information (carrier status and predisposition to disease) are highly private and from a theoretical standpoint should be governed by these rules. Because the GPA's narrow definition does not include genetic information derived from family history and tests for gene products, some genetic information that is private will not be protected. However, to more broadly define private genetic information to include such information (as the GCNA does in its definition of "genetic information") presents a vexing practical problem: how to distinguish between such genetic information and other medical information? Notations in medical records of the private genetic information subject to these rules will inevitably necessitate some changes to record keeping and disclosure procedures. However, if the information subject to these new rules includes family history and blood test results, virtually all medical record keeping would be affected. In fact, an early draft of the GPA incorporated genetic information derived from family medical history into the definition of "private genetic information." Because of the practical consequences of defining private genetic information in such a broad manner, we opted for the narrower definition and protection of the most private genetic information— that which is discovered by analysis of DNA.

Some have suggested that developments in molecular medicine which lead to routine use of tests for gene products, indicate a broader definition is needed, one that at least includes the results of tests for gene products. If such a test were conducted on an asymptomatic individual for the purpose of determining that person's future risks of developing a genetic disease or medical condition, it would have similar implications for individual privacy as analysis of DNA itself.[15] Blood tests to determine the presence of apolipoprotein E (ApoE) illustrate how the presence of gene products is used to predict susceptibility to heart attacks and may also be used as an indication of risk for Alzheimer disease, a late onset genetic disease.[16] As more is known about the associations between specific gene products and various diseases or conditions, analyzing DNA for genetic markers for those diseases may become unnecessary to generate the type of private genetic information that the GPA seeks to protect. This development would require us to decide in the future whether to treat the information from such tests the same as the information that is currently derived from DNA analysis.

## Fair Information Practices

Once the difficult hurdle of defining private genetic information is cleared, establishing information practices that will adequately protect individual interests is a less formidable task. The GPA achieves this goal by requiring those who maintain the information in the ordinary course of business (rather than friends or relatives with whom the person might share this information) to adhere to specific informational practices. Those provisions: require written authorization by the individual or legal representative before disclosures can be made,[17] limit disclosures of the specific information described in the authorization to those named in the authorization and for the purpose noted in the authorization,[18] grant individuals the right to inspect and obtain copies of records containing their private genetic information.[19]

Consistent with the notion that disclosures of private genetic information can have consequences for the individual that may differ in kind or severity than disclosures of other medical information, a general authorization for release of medical records is not sufficient for release of private genetic information.[20] Disclosure of private genetic information without prior authorization can only be compelled in limited circumstances, such as suits in which the genetic condition of a party is at issue or when the person maintaining the information is the subject of a criminal investigation in regard to such activity.[21]

## Minors and Incompetent Persons

Specific rules are needed to address special circumstances and the interests of particular populations that will be affected as genetic testing becomes more prevalent. Consequently, the GPA has sections governing the collection and analysis of samples from minors and incompetent persons.[22] It prohibits DNA analysis of children under the age of 16, even with parental authorization, for any condition that will not be manifest until after the child reaches adulthood, unless there is some effective measure that can be taken before that time to prevent or ameliorate the condition.[23] However, DNA analysis of asymptomatic children under 18 for research may be authorized by parents, provided that access to the results of the child's genetic information is held from parents if it reveals a condition that cannot be ameliorated, prevented, or treated while the child is under 18.[24] These somewhat unusual steps of restricting parental discretion are to protect the privacy interest of the adult the child will become.

Similar restrictions apply to the collection and analysis of DNA of incompetent persons and the dissemination of their private genetic information. Such analysis is permissible only when authorized by the person's legal representative, and when conducted for one of three purposes: the diagnosis of the cause of incompetency, the diagnosis of a genetic condition that can be effectively prevented, ameliorated or treated, or the diagnosis of such a genetic condition in certain relatives.[25] Disclosures of private genetic information are likewise restricted to circumstances involving such diagnoses.

**Other Special Sources of DNA**

Genetic testing may be most useful and most sought after in regard to reproductive decisions. Regardless of her age, the GPA gives the pregnant woman decision making authority over her own genetic analysis and that of her fetus, and authority regarding disclosures of the information that is created as the result of such analyses.[26] Having such information may be critical in regard to decisions related to pregnancy. These pregnancy provisions have been deleted from the GCNA. Similarly, obtaining genetic information about embryos that result from in vitro fertilization may be critical for decisions regarding implantation of such embryos. To facilitate such decision making and clarify who can authorize genetic analysis or the disclosure of genetic information regarding extracorporeal embryos, the GPA places decision making authority with the person or persons who intend to use such embryos for reproduction because they have the most interest in this information.[27]

**Conclusion**

Whatever the ultimate medical benefits of decoding the entire human genome are, the private information obtained from DNA analysis will have a profound impact on how individuals view themselves and others. The GPA is an attempt to establish rules for the protection of individual privacy as curiosity about perhaps the most private and sensitive information—genetic information—is driven by the piece by piece decoding of the genome. Since it focuses on privacy and confidentiality of information, it does not address all the policy issues involved in the creation of genetic information. For instance, prohibitions against use of genetic information so as to avoid harm to other individual interests are not included, but have been addressed in other proposals, such as the GCNA, on both state and federal level. We support legislative action to regulate the *use* of genetic information, and the GPA is compatible with these proposals. For all such proposals, however, the key problem is defining the information that is to be regulated. The GPA presents one definition as part of its comprehensive rules and has already had an impact. This definition and the GPA itself has stimulated debate as to how we can best protect genetic privacy, and stimulating and enriching this debate was one major reason for drafting the GPA. In that regard, the GPA is already a success.[28]

## November, 1997, Update on the GPA:

Despite indications that S. 1898, the Genetic Confidentiality and Nondiscrimination Act of 1996, had Republican leadership support, in June 1996, President Clinton called for enactment of S.89 (and its House companion H.R. 306), the Genetic Information Nondiscrimination in Health Insurance Act of 1997. As its title suggests, S. 89 narrowly focuses on the uses of genetic information in health insurance and ignores the much broader and equally urgent issues addressed by S. 1898.

## References

1. Annas, G. J., L. H. Glantz, P. Roche. 1995. The Genetic Privacy Act and Commentary, Boston University School of Public Health, Boston, MA (available by request from the Health Law Dept., 80 East Concord St., Boston, MA 02118, and also at http://www.busph.bu.edu/Depts/Health Law), a report that was funded by the Ethical, Legal & Social Implications of the Human Genome Project, Office of Energy Research, U.S. Dept. of Energy No. DE-F602-93ER 61626 with additional support by Boston University School of Public Health. *See generally,* Annas, G. J., L. H. Glantz, P. A. Roche. Drafting the Genetic Privacy Act: Science, Policy and Practical Considerations, 1995. 23 J. Law, Med. & Ethics 360-366.
2. Goleman, D. , Forget Money: Nothing Can Buy Happiness, Some Researchers Say, *N. Y. Times,* July 16, 1996, Sec. C, p.l.
3. Annas, G. J., and S. Elias, eds. 1992. Gene Mapping: Using Law and Ethics as Guides. New York: Oxford University Press, p.9 and *see,* Annas, G. J. 1993. Privacy rules for DNA databanks: Protecting coded future diaries, 270 *JAMA* 2346).
4. GPA Sec. 122.
5. GPA Sec. 121.
6. GPA Sec. 123.
7. An identifiable sample is one that is linked to any individual identifier "such as name, address, Social Security number, health insurance identification number or similar information by which the identity of a sample source can be determined with reasonable accuracy, either directly or by reference to other available information." GPA Sec. 3 (h) and (i).
8. GPA Sec. 101(b).
9. GPA Sec. 105.

10. GPA Sec. 103(a).
11. GPA Sec. 104(a) and (b).
12. GPA Sec. 103(c).
13. GPA Sec. 162.
14. GPA Sec. 3(m).
15. This is the trend in recent proposals on the state level. See for example Or. S.B. 276, 68th Legis. Assembly (1995). This also would be more consistent with the concern of the Task Force on Genetic Testing which focused on the implications for genetic testing for clinical purposes. Their broad definition of genetic tests includes the "analysis of human DNA, RNA, chromosomes, proteins or other gene products to detect disease-related genotypes, mutations, phenotypes, or karyotype for clinical purposes." It does not include family history despite the Task Force's acknowledgment that it can be an important screening tool. Interim Document for Public Comment from the Task Force on Genetic Testing of the NIH-DOE Working Group on Ethical, Legal and Social Implications of Human Genome Research (available from Task Force on Genetic Testing, 550 N. Broadway, Suite 511, Baltimore, MD 21205).
16. Mayeux R., and N. Schupf. 1995. Apolipoprotein E and Alzheimer's disease: the implications of progress in molecular medicine. Am. J. Public Health 85:1280-1284.
17. GPA Sec. 111.
18. GPA Sec. 112.
19. GPA Sec. 113.
20. GPA Sec. 112(f).
21. GPA Sec. 115.
22. GPA Secs. 141-144.
23. GPA Sec. 141.
24. GPA Sec. 131(c).
25. GPA Sec. 142(a).
26. GPA Sec. 151.
27. GPA Sec. 153.
28. *See, e.g.,* in addition to commentary in this issue, Holtzman, N. A. 1995. The attempt to pass the Genetic Privacy Act in Maryland, 23 J. Law, Med. & Ethics 367-70; Parmet, W. E. 1995. Legislating privacy: The HIV experience, 23 J. Law, Med. & Ethics 371-374; Clayton, E. W. 1995. Why the use of anonymous samples matters, 22 J. Law, Med. & Ethics 375-77; Reilly, P. 1995. The impact of the Genetic Privacy Act on medicine, 23 J. Law, Med. & Ethics 37881; Lin, M. M. J. 1996. Conferring a federal property right in genetic material: Stepping into the future with the Genetic Privacy Act, 22 Am. J. Law & Med. 109-134.

# THE BRAVE NEW WORLD OF DNA: TRANSCENDENCE OR TRANSGRESSION?

## Vijaya L. Melnick

A myriad of perspectives on the subject of the Human Genome Project, genetic testing and screening, and some of the consequences have been presented. My paper is titled "The Brave New World of DNA: Transcendence or Transgression?" In an ideal world we would expect that the knowledge that we encounter through research would produce the best results for each one of us and society as a whole. That would prove to be the transcendence of all our pain and suffering and lead us toward a utopian ideal where the welfare and well being of all is the goal. But ours is not an ideal world. Ours is a very stratified, unequal, and fragmented society, and what becomes of the knowledge in that society and how that knowledge permeates society in various places are of great concern.

It has been said by many intellectuals that knowledge is good, knowledge is objective, knowledge is neutral. However, knowledge benefits—as we have seen far too often—people who already have a lot of knowledge. It very seldom benefits people who do not have knowledge or access to knowledge. In fact, it very often creates a larger chasm between the "haves" and the "have nots." When current medical practices deny some of the best technology and best procedures to some people, why would we expect that genetic technologies will be accessible on an equal basis? Logically, it does not follow. Therefore, what I will try to do—and what we have not done so far in this very comprehensive symposium—is to look at the question from the point of view of people who have a "genetic disease."

We have heard about surveys. Surveys are said to be faceless. The tears have been wiped away. We have heard a vignette from a panelist about a person who has struggled with the knowledge that she may have the breast cancer gene. But what we have not heard, what would be very useful for all of us is to

The author is Professor of Biological and Environmental Sciences and Senior Research Scholar at the Center for Applied Research and Urban Policy at the University of the District of Columbia, 4200 Connecticut Ave., N.W., Washington, D.C. 20008.

listen to a person who has one of these so-called genetic diseases or genetic disorders. We need to listen to what they think, or what they expect of their own lives, whether they think that their lives are futile or defective or disabling, and in what sense.

I think our very definitions of disease, disability and disorder are based on our own concepts, and the concepts of a society which does not easily tolerate differences and is not very friendly to people who are disabled. What I tried to do was glean from the points of view of people who have lived their lives with just such disabilities. I would recommend that you read, if you have not already done so, the book by Oliver Sacks entitled, *An Anthropologist on Mars (1995)*. He writes about six or seven case histories about people who have various disorders and how their lives have impacted society. He tells about the tremendous contributions they have made and asks whether we should consider them abnormal. In my opinion, anyone who reads that book and comes to that conclusion would be the abnormal one.

I think our concepts of normality, abnormality, illness, disease and disorder must be rethought. Because, if you are talking about a cure for a genetic disease, you are essentially talking about eliminating it. People are very concerned, especially those with these genetic disorders, that an attempt will be made to eliminate those who have such a disorder.

Instead of trying to see how we can live in a society with people of varying talents, it seems that we are trying to see whether we can knock down differences and have a very homogenous, uninteresting place where people are all going to be the same. I think that is going to be very boring.

Oliver Sacks makes this statement in his book. "I am sometimes moved to wonder whether it may not be necessary to redefine our concepts of health and disease, to see these in terms of the ability of the organism to create a new organization and order, one that fits its special altered dispositions and needs rather than in terms of a rigidly-defined norm."

Kay Jamison, a professor of psychiatry at Johns Hopkins University, has written very eloquently about the manic depressive disease that she has suffered all through her life. Her book, *The Unquiet Mind* (1995), is a great contribution to society. She tells about her life-long struggle with this disease and how she finally learned to live with it. She also talks about the artists who have lived their lives with such diseases.

In fact, creativity and depressive diseases often have been twins. Therefore, if we are trying to eliminate manic depressive diseases or mental disorders rather than help people with the disorders learn to manage them, we will be doing away with a whole lot of creativity that we have experienced. You can go down the list of artists and see how many of them were people who suffered from such diseases and, if they had not lived, this world would not have been as pleasant a place. Jamison writes, "If we got rid of all the manic depressives on medical school faculties, not only would we have a much smaller faculty, it would also be a far more boring one."

The other thing I would like to mention is that we do not see studies of people in medical schools or law schools or in corporate America who have such diseases and what they are doing with them. We see that these kinds of studies are mostly done in poor communities. They look at how much disease is prevalent in those communities, and what we should do about it. And they are conducted by those who wield authority.

The definition of an "acceptable disease" in these times is one that afflicts people who have the possibilities of dealing with it. Some people think it fashionable to say, "I'm on Prozac." It is as if it confers status on them. But, a poor person saying that on a job application form would be looked upon quizzically as if he had a real problem. He would not get the job. The way we look at people with the very same illness is dependent on their status in this society.

Sacks further comments that "nature's richness is to be studied in the phenomena of health and disease, in the endless forms of individual adaptations by the human organisms, people, adapt [sic] and reconstruct themselves, faced with the challenges and vicissitudes of life. Defects, disorders, and disease in this sense can play a paradoxical role by bringing out latent powers, developments, evolutions, forms of life, that might never be seen or even be imaginable in their absence."

It is a paradox of disease that it sometimes is the very instrument which brings out the creative potential of many human beings. We really need to learn how we accept that challenge from people who have already accepted the challenge. Very often they have much more fortitude; they have much more courage, and they have many more talents than the "normal ones of us," who might not have been able to deal with them. Are we prepared to sacrifice that courage, those talents, because they have one problem that they could manage with the help of all of us? This is the question that we have to answer.

An animal science professor, Dr. Temple Grandin, who is highly regarded in the veterinary sciences and is autistic—says, "If I could snap my fingers and be non-autistic, I would not. I would not because then I would not be me. Autism is a part of who I am." Because she believes that autism may also be associated with something of value, she is alarmed at thoughts of eradicating it. In a 1990 article she wrote, "Parents are often angry with autism. They may ask why nature or God created such horrible conditions as autism, depression and schizophrenia. However, if genes that cause these conditions were eliminated, there might be a terrible price to pay. It is possible that people having bits of these traits are more creative and perhaps even geniuses. If scientists eliminate all these genes, maybe the whole world would be taken over by accountants!"

Talking about herself she says, "I have heard that libraries are where immortality lies. I do not want my thoughts to die with me. I want to have done something. I am not interested in power or piles of money. I want to leave something behind. I want to make a positive contribution and know that my life has meaning. I am talking now about things at the very core of my existence."

Who are we to say that this person is less than any one of us? We cannot, and therefore I fully believe that when you talk about diseases that you plan to cure, that you plan to eliminate, that you remember that behind each may be a human being who is far better than you or me or a whole lot of us. We have to think carefully, and try to create a network, a society where people are supportive of each other and where we can look on each other as a talent, or a help or a resource that we can tap into. Let us not in our haste and in our short-sightedness eliminate things without knowing in the long run we may be eliminating ourselves.

## References

Grandin, Temple. 1990. Needs of high functioning teenagers and adults with autism. Focus on Autistic Behavior 5(1):1-16.

Jamison, Kay Redfield. 1995. An Unquiet Mind. Alfred A. Knopf, New York, NY.

Sacks, Oliver. 1995. An Anthropologist on Mars. Alfred A. Knopf, New York, NY.

# WHO'S THAT FISH ON MY LINE?
## —The Dangers of Electronic Distribution of Genetic Information

### Hans S. Goerl

The revolution in human molecular genetics and the revolution in electronic communications each has great potential for good and evil. The fact that they are occurring simultaneously heightens both of those potentials. When one bears in mind the cardinal principle of the Hippocratic oath—first do no harm, it is apparent that not enough attention has been paid to the potential for evil that the immediate electronic availability of genetic research and genetic information presents.

The beneficial potential is quite obvious—rapid distribution and availability of huge volumes of new information. The dangers, however, are subtler and generally unappreciated.

Very few physicians or scientists will dispute that the field of medicine which has seen the greatest expansion of knowledge in the last five years has been human genetics. The Human Genome Project and private investment have pumped hundreds of millions of dollars into genetic research and this investment is now resulting in the clinical or diagnostic application of genetics in every discipline from oncology to sports medicine.

Historically, new medical information and techniques have passed incrementally into general medical practice. First came many years of obscure laboratory work, then initial publication in specialized scientific journals and then in "clinical" journals such as *The Lancet, JAMA* or the *New England Journal of Medicine*. With the exception of a few special cases such as AIDS, the public has remained fairly ignorant of new "not ready for prime time" information and techniques until the medical establishment has had time to evaluate and test them in clinical settings. Most patients who might have benefitted from this new knowledge have remained in blessed ignorance of the latest developments;

A practicing attorney, the author is the founder and executive director of The Genethics Center (21 Summit Avenue, Hagerstown, MD 21740), an advocacy group dedicated to education about and opposition to genetic discrimination. He is also the Ethical and Social Issues editor of Hum-Molgen, a listserve for geneticists which has over 4,500 subscribers in 64 countries. Hum-Molgen can be accessed at http://www.informatik.uni-rostock.de/hum-molgen ©May not be reproduced in any form without the expressed written consent of the author.

those who sought to discover the limits of what their doctor knew or was willing to tell them were often subject to subtle stigmatization as "difficult" patients or "doctor shoppers."

A variety of administrative hurdles, most notably the FDA in the case of drugs and devices, have also retarded the progress of the latest tests and techniques into the clinical realm. Now, however, for a variety of reasons, the pathway from the lab bench to clinical application and from the lab notebook to the headlines have both dramatically shortened. Just to give one extraordinary example of this, less than one year after the *BRCA1* gene thought to predispose people to breast cancer was identified, a clinical test for the gene was being marketed nationwide and in many other countries.

Genes thought to predispose to different diseases are identified daily and headline grabbing gene-therapy experiments are announced almost weekly. The widespread popular perception is that genetics is on the verge of curing most of what ails us.

It is little wonder that millions of people are now taking to the electronic highways to find out the latest about "their" or "their family member's" disease. Many of them are facing immediate grave crises in their lives. Many others are scared because someone else in their family has a disease now thought to be "genetic," and they are afraid that they or their children are next. Almost by definition, they are uniquely vulnerable.

They are using Usenet newsgroups, electronic mailing lists, web sites, chat rooms, medical forums and powerful net search engines to get information that, only a couple of years ago, was almost completely inaccessible to them.

Anybody with a computer and a modem can use Online Mendelian Inheritance in Man (OMIM), or read the home page of a researcher working on "their" disease. It is only a matter of time before somebody makes available genetic cancer predicting software similar to that which several biotech companies are using. When one considers that around 40% of practicing physicians have never had a single course in genetics and that there are only 2,000 or so qualified genetic counselors in the entire country, it is clear that there will be, in many cases, nobody to explain this information.

Unfortunately, none of the circumstantial guarantees of trustworthiness that surround traditional printed, broadcast or spoken communication are present on the Net. For many Net resources, there are no peer reviewers, there are no editors and there are no proofreaders. There is nobody to modify the information or to customize its presentation so that it is understandable or appropriate for the reader's age, mental condition or particular vulnerabilities.

There is not any guarantee that the purported source is the real source. When you sign on to the Megadrug company's web site, are you really sure that it is Megadrug's site? When Yahoo's search engine directs you to the "National African-American Medical Committee" web site for sickle cell disease, and it tells you that grandchildren of carriers should be tested, where did that information come from? For that matter, what is the NAAMC? Is it real? Or is it a front for a drug company marketing its tests?

Although it comes to you via the most modern of means, there is often no way to discern if the information is still accurate. If the same site tells you that the alpha-oxy-one test is not accurate in people of Sudanese descent, when was that information written—last week, or five years ago, before the alpha-oxy-two test came out and just before the Committee stopped updating its website?

There is also no way to tell who on the Net is interested in you. When you checked out that page on the web-site, who noticed? Is your name now on a mailing list of likely sickle cell sufferers. Does your insurance company buy that mailing list?

The anonymity of the Net can be dangerous on other levels. When you post a message to a Compuserve forum asking for information about the genetics of diabetes, and Dr. Whitecoat answers, how do you know he is a doctor? There are lots of wannabes out there.

And finally, even when you finally find information from an "unimpeachable" source such as a major research institution, are you sure they do not have a hidden agenda? Do they receive patent royalties from the genetic tests they write about? Are they hoping to discover genetic cohorts or families whom they can invite to participate in research projects? Do they run their own HMO, which might be interested in your genetic information?

Some of these problems are solvable. At least one organization has proposed a voluntary code for all medically oriented web-sites. If the publishers agree to abide by the code, they get the right to display the organization's seal of approval.

The editing of Usenet newsgroups is a vital function that should be underwritten by the scientific community. At Hum-Molgen, which has six editors on four continents, virtually every day we receive patient requests which divulge intimate genetic information. If these requests were not edited or re-directed, they would go out to all 4,000 of our subscribers as would the responses. But such editing takes time and effort.

Massive efforts to educate the public about the uncertainties of the scientific process would also help as would a movement toward greater publication and dissemination of negative experimental results.

But in the final analysis, there is no way to truly control the information revolution. In terms of societal goals, that is probably a good thing. Therefore, the best we can hope to do is to encourage self-restraint and sensitivity on the part of information providers and a healthy sense of skepticism and caution on the part of net users.

# 4

## Recapitulation

# OF GENES AND GENOMES

## David Botstein

My original plan was to start with the science policy issues and the history of the Human Genome Project and how things came to be. But, since the ELSI (ethical, legal, social issues) tide has washed over this meeting to a very considerable extent, I better say a few things about some of the issues of a scientific nature which I think ought to serve as boundary conditions for the prurient speculation that has a tendency to excite everybody's ELSI instincts. So first we will review the history and then we will go on to the prurient things.

I want to make some points about the goals which the Human Genome Project set for itself back in 1990 under the stewardship of Jim Watson. First of all, some of them have clearly been achieved. We are beyond the first goal of constructing a high-resolution genetic map of the human genome. We can definitely see the end in sight for the second goal, the sequencing of human DNA. All but the shouting is over in terms of giving out the money for sequencing, and the main players are well defined. The technology is well defined, but the sequence is not yet in hand. Regarding the selected model organisms to be mapped and sequenced, we have sequences for three species of bacteria, and the complete sequence of the first eukaryotic organism, *Saccharomyces cerevisiae* (brewer's yeast) was announced in April, 1996. A great deal has been learned.

The second thing that I would like to say is that what we have today is the result of a compromise between extreme views. One of those views was that we should simply sequence the human genome as quickly as possible using the available technology in 1989 no matter what it costs, and the other extreme view was that sequencing of the human genome should not be done at all, and we should just wait for nature to run its course. The compromise differs from both of the views in that it has provided us a with a large amount of useful information which the genome insiders are now pleased to call "infrastructure." In fact, I think it is an appropriate term because we now have a system of roads on which an industry of genomic trucks can ride efficiently and provide all

The author is Professor and Chair of the Department of Genetics at the Stanford University School of Medicine, Stanford, CA 94305-5120.

kinds of useful information. The most obvious useful information will be the identification of disease genes on a wholesale basis instead of on a retail basis.

Francis Collins, director of the National Human Genome Research Institute of NIH, made a realistic estimate when he projected it would cost in the order of many tens of millions of dollars to find the cystic fibrosis gene. He has been variously quoted as saying anywhere between $50 to $200 million. Currently, an equivalent job with equivalent family resources using the infrastructure that now exists is a work for a talented graduate student in his or her Ph.D. career. That represents a significant reduction in the cost for the same job. Not bad for scientific policy. I might add that it is a successful scientific policy. It is my opinion that we need to remind ourselves about what has been achieved and what has not been achieved. Then I want to go a little bit into the limits of what can be achieved.

In my opinion, things like creating the "perfect baby" and "enhanced genes" will not happen because they cannot be done for well understood reasons. I call these kinds of things "bunkum." They are not going to happen. On *Star Trek* they may do these kinds of things, but they will not be done in real time. So our worries should be about more realistic things. Racism is a real thing. People can be injured, people can be offended, and people can be abused. It is useful to have a boundary on what is expected to happen realistically.

Let us look at the science. In the $3 \times 10^9$ base pairs in the chromosomes is all the information which we can think of as inherited. There is much discussion currently about what is heritable and what is cultural. From the scientist's point of view, it seems that, if you cannot distinguish anything except that the children are like the parents, then a distinction cannot be made between genetic and environmental influences. If, however, a robust relationship can be found between the inheritance of a particular piece of DNA or several pieces of DNA from the genome in only those people who have a particular trait, then there is an unbiased, robust argument in the scientific sense that says it is likely—and if the numbers get large enough, then it is virtually certain—that this is an inherited trait, at least in part. That is all it is. It is nothing more than that. But that is a lot! Therefore, one can associate the transmission of a trait with a piece of DNA.

By using polymorphic markers one can obtain detailed information about which children get which part of which chromosome from which parent. This is currently being done on a wide-scale and is easy to do. If this is done on traits that cannot be measured, then results will be inconclusive. But when done on measurable traits, the position of markers all over the genome can be viewed and then the likelihood of correlation examined. There are mathematical limits to how well we can do this procedure.

DNA markers can be followed in families, and this allows us to find genes. Some of the genes that have been found are cancer genes. In fact, there are many such genes—thousands—that have been found at the linkage level and a few, 100 to 200, have been cloned. Getting to where we are today required a fair amount of new technology and, at the time the NRC Committee wrote the specifications for the compromise, many of the current pieces of technology were not available. They were anticipatable but not anticipated exactly. Others were unknown at the time. In particular, the size of the genome was incommensurate with the means that were being proposed for sequencing it. There is a lot, but not an infinite amount, of information. In fact, most people probably have a computer that can remember the human genome—although it cannot understand it.

After genetic mapping is completed, there is a level of resolution where the nearest markers are indistinguishable on the physical map. When the physical mapping (infrastructure) is completed, a DNA sequence—one that for the most part no one understands—is obtained. Most of the DNA sequences that are derived do not mean anything to the scientist.

How is the map made? When we first began, we did not have a good method, but Maynard Olson figured out that maps could be made using large vectors. He also figured out how to use PCR to make unique markers which could be ordered. In fact, the method of ordering, also not seriously considered by the NRC Committee, was an invention of David Cox and Rick Myers. They figured out how radiation hybrid mapping together with STSs could be used to map in a very efficient manner. This new technology emerged after the project got started.

It is now literally the case that a scientist who has a piece of DNA and wants to know where it is located on the genome does the following: he or she takes the piece of DNA, produces a unique sequence, makes STS primers and puts them on the order of 50 to 100 radiation hybrids, does the PCR, that is 50 PCRs—a morning's work, then sends the results off to a computer at Stanford or at the Whitehead Institute and usually within an hour gets information about where the piece of DNA is within considerably less than 1% of the genome. This is not bad for five years of work.

At the time of initial discussions, no one could have imagined a system that was so quick, so easy or so cheap. Also not anticipated was the growth of a new industry as a result of this work. Research Genetics was the first company. Instead of having a warehouse of probes and PCR primers that he was giving out to people while being accused of not making them available, Eric Lander decided to convince a businessman to do it for him and to make money at it the American way. So by doing this in large enough volumes and at modest costs, the scientific community can now afford to do these experiments on a massive scale. This was not planned. It happened.

Once you have physical maps and meiotic maps, the sequencing of the DNA can be done. What does that have to do with traits? It means that, if you have a trait that can be measured accurately, it can be mapped if—and only if—a whole set of conditions applies.

There are many different types of limits. Eric Lander and I wrote a paper in 1986, which has still not been refuted, in which we claim, based on mathematical arguments alone, that even if I have a trait which I can unambiguously measure like height or color of skin and even if it is inherited, i.e., by hypothesis, you know that the genes are responsible for all of it, there are limits to what one can figure out. I do not actually know any such real trait. This is all in the realm of mathematics, but that is the best case. If there is any ambiguity about the trait, of course, things get worse, not better. So in the best case you can ask the question—how many genes can I figure out by these methods, using a finite number of subjects, let's say, the population of Europe? The answer is: I cannot figure out very much; there is a limit.

The practical limit that we considered—which was hundreds of families with three affected individuals—was three or four genes. Anything that is more complicated than that we are not going to figure out, even if we have the entire sequence of the human genome, even if we genotype every living human on the earth everywhere. This is not the Human Genome Project. This is the everybody project, but I still will not be able to figure out things that are more complicated.

I should tell you that these kinds of limits are not unusual in science. For example, you all know about quantum mechanics, the atom bomb and all the rest of that. I should tell you that quantum mechanics and these other things work pretty well. We can calculate pretty exactly where all the electron density is in a hydrogen atom. With a lot of work we can do the lithium atom, but there it stops. No one can do better than that. After that we are talking approximations and the approximations get extremely lousy as you go up in the Periodic Table.

We are never going to do that, because the problem is not one simply of computation. There are mathematical problems having to do with the specifications of relationships of bodies with each other. How many bodies can we follow? Fewer than we can figure out genes in a population. These are mathematical limits. They are not going to go away. So if it is more complicated, you are not going to figure it out.

The second problem is high frequency. All this linkage stuff is based on an assumption that the frequency of what you are looking for is relatively small. If the frequency is low, then rare diseases are a piece of cake. This is what all these graduate students are mapping. However, if things are really complicated and there are many genes involved, and they are common—like manic depression, hypertension, schizophrenia—then we cannot figure them out because they are not rare enough. Here is a calculation I made on the back of the envelope this afternoon: If the frequency of disease is about 2% and five genes contribute equally as dominant factors, each one has a frequency of 0.46. That means that half of the population has the wrong one of each of these factors. When you get all five of them together, then you are in trouble. That is not a solvable problem. There is no resolution there. There is no power to discover anything. We are not going to figure this out.

Finally, I come to probably the most serious problem of all for the most popular of your concerns in the ELSI world, and maybe in the lay world also, and that has to do with the definition of a phenotype. Let me take an extreme example: you want to find out if there are pieces of DNA associated with goodness. That is, you are a good person if you have these markers, and you are a bad person if you have alternative alleles. In addition to many other considerations, because there are lots of good people in the world so that the frequency has got to be high, it is so complex it is not going to work for that reason. However, it is not going to work for an even more profound reason, the more profound reason being that we cannot tell who is "good" in any convincing way. What are we going to do? Give a lie detector test? Look at people's CVs? What do we mean? The problem, in fact, is the same for all kinds of traits going down the list of things that you might worry about like IQ (whatever that means!) or speed of running.

Actually I have a ditty that I tell. I have been telling my classes this for 20 years, and I teach this to every genetics class I ever give. I thought maybe this audience ought to hear this argument. I call the argument a *gedanken* experiment, which is an experiment you do not have to do. Just by designing it and thinking about it, even if it has elements that cannot actually be carried out in practice, you achieve an understanding of a principle that makes you not want to get involved in any such thing. The most famous example is the one that Einstein did of thinking of himself as an electron on an elevator. My example has to do with genetics and basically the experiment goes as follows:

We are going to look at the genetics of running in human individuals, actually nine-year olds, and in our *gedanken* experiment, we find a town in the rural Midwest in which there are exactly 100 nine-year-olds, and from the nine-year-olds and their parents, we get informed consent to have a 100-yard dash on a nice Sunday afternoon. It is an ordinary Sunday afternoon—no thunderstorms, the pressure is normal. The whole town comes out to see the 100-yard dash. We line the kids up at one end of the field, and at the other end we sit with our stop watch and we clock them as they come by. It is a small town and everybody knows which kid is which; there is no confusion about how long it takes to get the 100 yards.

Then, as statisticians would do, we plot the results. On the plot we show number of kids and time. In this plot, the left side is "fast" and the right side is "slow," —fast means short time, slow means long time. What does this distribution look like? The main body is our favorite normal distribution, and I will argue in our thought experiment that in fact there is a little bump over at this end and a little tail over at the other end. The tail are kids, one or two in 100 probably, who have serious physical problems. They may have a club foot, and they are going to do the 100-yard dash slower. The bump on the other side are the two or three kids in this town who are the offspring of fitness nuts. These parents are out running the 100-yard dash every morning and every evening, and the kids have been running along and mimicking the parents. They know all the techniques. They know how to move their arms, crouch, listen to the gun. They do all these things that make you run the 100-yard dash faster. They have technique, and there are two or three of them.

The rest are your basic normal kids and they run the 100-yard dash in typical fashion. We know as a matter of many generations of experience, that a nine-year-old without prior training will run the 100-yard dash in minutes. We also know that the world record is something under 10 seconds. There is a long way between these two distributions.

The rest of our *gedanken* experiment goes as follows. We hire a coach from Romania; we bring this coach to our town. Everybody including the coach is teaching these kids, as best we can, to run the 100-yard dash. He does it; they do it, and everybody is happy. The kids are as fit as can be; they all know how to do everything and this will be the town's source of Olympic glory. On the anniversary of the first run, we run the 100-yard dash again. The kids are all 10 years old, but that's all right because there is internal control. What do you think happens?

The new distribution would show that the kid with the club foot is ahead of where he was but still behind, and everybody else is now sitting on top of the fitness nuts. Whatever advantage they had is gone; they are all moved into Olympic territory—a few seconds for the 100-yard dash, not a minute. So far everybody wins.

Could I say to you, which I won't, that the distribution on the left is entirely the result of genetics plus random chance? In order to do this, we would have to run the 100-yard dash many, many times, make the average of averages and get the distribution. But if the rank order in that distribution was entirely the result of genes, even then would I be able to predict the position of any individual kid (except the one with the disability) in the first run? The answer is logically, physically and mathematically: no, I cannot. The variance here, the variance there and the distance between the means makes it vanishingly unlikely that I could. Too many other variables. It is not going to happen.

Is this different from the situation with IQ? No, it is not. Anything that is trainable—in which the mean moves some substantial distance for unknown reasons—cannot be figured out. You can sequence your little heart out, you can calculate, you can run the 100-yard dash, you can do whatever you like, but you are not going to figure it out. Not from this distribution.

So from a scientific point of view, this whole business of *The Bell Curve*, atrocious though the claims may be, is nonsense and is not to be taken seriously. People keep asking me why I do not rebut *The Bell Curve*. The answer is because it is so stupid it is not rebuttable. You have to remember that the Nazis who exterminated most of my immediate family did that on a genetic basis, but it was false. Geneticists in Germany knew it was false. The danger is not from the truth, the danger is from the falsehood. That is my ELSI piece.

Let me get to the good which is going to come from the Human Genome Project. Here again we are unfortunately at the mercy of the press and the prurient and so on. What good is going to come out it? Well, people are hard-pressed to say. Every time we say something obvious, someone says something about informed consent or something or there will be psychological damage. I think it trivializes what we are about, because there is really major league good about to come, and it is not gene therapy.

I happen to be among those who believe that gene therapy is a real possibility, but it is pretty far off. I can think of two or three technical hurdles that are much more serious than the ones we said we would solve in 15 years. So using that metric, I think that we are 20 or 30 years away from wholesale use of gene therapy for any kind of useful purpose to actually cure disease.

So if it is not gene therapy, what is it? Let me paraphrase a talk that the new head of the Cancer Institute, Rick Klausner, gave to a bunch of genome gurus. I will take a few liberties with the data he presented which I do not have with me but they would show you a survival curve of the ordinary epidemiological type for a cancer—a bad one. What you would see would be that, at time zero, 100% of the people are alive and in a year or two they would all be gone; they died of the cancer. This is with no treatment; bad business for everybody.

But we could draw a different picture showing two different treatments for which clinical trials have been carried out: Treatment A and Treatment B. You see that there is good and bad news. With Treatment A people die more slowly and some of them survive. With Treatment B they die faster and some of them survive. So the FDA and the Cancer Institute go around choosing between these treatments, and in fact they have to choose between them. Your oncologist, if you have the misfortune to have this cancer, is going to sit around looking at these data and say: well, what do we want? Treatment A or Treatment B? He and you will probably choose Treatment A. Why? Because you live a little longer. This is the state of cancer treatment today.

Here comes the fun. By genetics it will become possible, and I do not believe that this is unlikely—I believe that this is a near certainty and so does Klausner and everybody else—that cancers might fall into two or more categories, but probably not 100 classes, for a particular organ type. Let's say, for the sake of argument, that they fall into two classes by genetics. So this is where we get informed consent and we go around and we take DNA samples, and we figure out whether we have *BRCA1* or *BRCA2* or something else. The chain of reasoning for those of you who do not remember is that there are some inherited cancers in which victims have a head start on cancer because what they would normally get by a series of events, they have inherited by a bad allele and so the cancer comes faster. It turns out that in many—although by no means all—of these cases, if you look at the sporadic cancers, the same genes are involved. That is what I mean by classification.

If Class 1 in these two classes is cured completely by Treatment A or B, and Class 2 is cured by Treatment A only, what do you think we get? We get out of these two treatments one system that completely cures the cancer. With Treatment A, the Class 2 cancers are cured. Nobody is dying. But the victims of the Class 1 cancers which are unaffected by that treatment are dying as before. With Treatment B, Class 1 victims are surviving, and Class 2 victims are dying as before. So, if I can tell in advance which one you have, you get the right treatment and you survive.

You may say this is an unreasonable scenario. No, this is not because in some tumors, for example, lymphomas—especially the juvenile ones, the moral equivalent of this has already been done. There are some sub-types of leukemias from which people generally walk away after treatment. Not all the treatment is pretty but survival is very good.

That is what we are after. This is not a minor point. Without the Human Genome Project, we can forget about doing this in our lifetime. That is what this is for, and so it grieves me a little bit to hear about the psychological harm because I am one of these guys who believes in the weighing of the good against the bad. I do not want to trample on anybody's individual rights, and I do not think that anything like that is required. But we should remember occasionally what the project is about. This is what it is about.

To conclude, there are going to be no designer babies and, if you ask me why not, it has to do with the large number of genes—as well as all the things I told you—and the small number of eggs, that is basically the outline of the answer. There are going to be no designer babies. Gene therapy someday, but diagnosis and subdivision of disease so that you can make appropriate cures— those are really around the corner.

# MOLECULAR HALOS AND BEHAVIORAL GLOWS

## Troy Duster

It is a very good thing when people come across disciplines and perspectives to engage each other in dialogue. They should do it in a spirit of collegiality and not in an adversarial way, unless the adversarial relationships produce some kind of emergent connection which makes things better. Maybe that will happen. Maybe we will see what have been called the "ELSI types" and the molecular genetics field really come together. I do want to emphasize the notion that this is not about good guys versus bad guys, about smart people on one side and dumb on the other, about clear versus fuzzy thinkers, or about people who are arrogant and those who are humble. On both sides there is arrogance and humility. On both sides there is fuzzy thinking. What I want to do is say something about fuzzy thinking on the science of race.

Let me start with a joke, not because it is funny but because it suits my purpose. It is about the relevance of this conference here in Tuskegee, and then I will switch to the connection between why it was brought here to Tuskegee, why the Human Genome Project has been kind of a floating game out there with very little connective tissue to the African-American community or to the experience of African-Americans. Very few people come to these meetings who are African-American. At least in the big picture, there is a remarkable bias toward race.

Here is the joke. There are three African-American college students in New York City just coming out of class, and one says to the other two: "I have just learned today in my biology class that race is of no use in terms of understanding neurophysiology, biochemistry, that we are all very much alike, that the variation between groups is not very important, that race is really not an issue in the biological sciences anymore." One of the other students, just coming out of an English class, said: "Well, I have just heard that, according to post-modern critique, race is only a master narrative which can be reconstructed and reframed, and so it is all just about a narrative account." The other student came out of a

The author is Professor of Sociology and the Director of the Institute of Social Change at the University of California, Berkeley, CA 94720.

class in social science and says: "I have learned that racism is socially constructed, that it is an epiphenomenon." Then they turned to each other and wondered why they could not get a taxi in New York City. The edge of the joke: if you read the newspapers in the last few weeks, you would have learned that Ed Bradley of CBS's popular *60 Minutes* sued a New York taxi driver. It appears that Bradley could not get a taxi, so he sued and successfully got a few hundred dollars. He made a big point of it, and it was a front page story in the *New York Times*. I will return to taxi drivers and the study of race in a bit because it has to do with the science issue and race.

Let me set the stage. Human genetics has made some real advances. One should not trivialize or underestimate the power of the diagnostic procedures in human genetics. Developments have been remarkable in the last three decades. I need not chronicle the vignettes, but I would say that David Botstein slipped over a point which I think needs a bit more emphasis. He talked about the short time frame in which there is a problem between diagnostic power and therapeutic intervention. I think he would acknowledge that this is a huge issue and, as the Human Genome Project continues in the next two to five years, what we will find is a remarkable change, that is, as diagnostic tools increase, therapeutics will stay relatively constant. We will find this gap increasing in the next five to 10 years to the point of it being a crisis. A crisis for whom? Certainly not for those who are involved in the field of diagnostics, molecular genetics or perhaps clinicians, but rather for people who have the information about which there is nothing more than the answer: "You are predisposed to (blank)," or "You have the gene for (blank)." There is diagnostic capacity out of this project, and it will be remarkably powerful.

There are also some implications for diagnosis. Once the information is available, it may be that you can alter the behavior of the patient and that is a good thing. So I want to echo that there are some good things coming out of the Human Genome Project. At the same time, in the last 20 years there has been a resurgence of claims for the genetics of complex behaviors. I want to draw the particular relevance of this to why African-Americans might have a special concern, and this in turn relates directly to why this conference was called.

Throughout the conference we have heard some rather confounding statements about the nature of race and science. It is kind of intuitive. Here we are in this setting, and we are told with great firmness, passion and conviction that, as in my joke with the three students, race is not a very salient category in the biological sciences. About a year ago, a consortium of leading scientists from biology, neurophysiology and physical anthropology—and several other disciplines—issued the revised UNESCO statement on race. It was a definitive declaration that summarizes 11 central issues and concludes that, in terms of scientific discourse, there is no such thing as race that has any scientific utility at least in the biological sciences.

The actual title of the UNESCO statement is what I want to talk about. The title is: "Is race a legitimate concept for science?" Note that it did not say biological science, it said "science." The UNESCO statement is ultimately about the problem of the difference between what I shall call, after one of my mentors, first order constructs and second order constructs in science. Some 50 years ago, Felix Kaufman (1958) made a distinction which throws some light on the controversy. Professor Jackson in her presentation said something like "We acknowledge that the biological concept of race has no utility in terms of the taxonomical system." But yet she says several times that "we need to be at the table." There is crackling tension in that formulation: it has no utility in terms of biology and yet she talks about these things in terms of their characteristics in the population and why we should "be at the table." You heard it from Drs. Cavalli-Sforza and Botstein, too.

Kaufman has some insights for us. He noted that there are several different kinds of issues, methodologies and theories that are generated by what he called "first order constructs" in the physical and natural sciences. For the physical and natural sciences, the naming of objects for investigation and inquiry for conceptualizing and for empirical regularities is in the hands of the scientists themselves. Here is the issue. On the one hand, there is the scientist and on the other is the empirical world of neurons, protons, molecules, DNA structures. The scientists construct ideas and concepts about the empirical world which are of the first order, and they are not mediated except by other scientists who can agree or disagree about the nature of the construction.

That is quite different from the task of the observer of the social world. That is because humans live in what is called a "pre-interpreted world." Humans grow up in families, cultures and societies in which categories are pre-existing so, when we arrive, we learn about concepts like family, crime, delinquency, love, truth, beauty, and concepts like race. So the task of the social observer of human behavior is to address, at least in Kaufman's insightful terms, the concepts that people actually have. That is, we must construct a second order of conceptualization. It is a fundamental mistake of social analysis to act as if the world is not pre-interpreted by those who live in it. To understand the world properly, the subjective, phenomenological world of those who treat each other with these concepts must be understood. Our concepts need to be attentive therefore to concepts of common sense actors.

Could this therefore mean that we cannot be "scientific." I thought science was about apprehending the empirical world with methods of investigation which are systematic and clear and based on theories, hypotheses, axioms and postulates. I do not want to get into a defense or attack of any discipline, but could one have a scientific study of taxi drivers in New York City and how they attend to the first order construction of race in their own world of pre-interpretation? Could we stratify a random sample across different hours. One could go to New York City with enough money and do a study over a period of about three months. One could use video cameras and observe how whites, Asians, blacks are treated on the streets of the city by taxi drivers included in a study controlled for class, income levels and regions of the city. In short, one could conduct a rigorous empirical study of taxi drivers in New York City, and how they pick up or do not pick up people in New York by the taxi drivers' conceptions of race. Would that be science? By any definition, that would have to be science.

The molecular genetics field over the last 30 years has made the most advances. On the one hand, we say that racism is not a construct of any use in biology, and then we say that gene frequencies demonstrated in the use of polymorphic markers occur more frequently in certain populations than in others. The distribution of those populations happens to coincide with racially preinterpreted groupings which we all can recognize, and thus the conundrum. For example, northern Europeans' greatest concentration of risk was for cystic fibrosis in this country, southern Europeans—especially from the Mediterranean—for beta thalassemia, and African-Americans for sickle cell anemia. Increasingly, we are finding out on the west coast that Laotians, Cambodians and Chinese are at greater risk for the other thalassemias. Can it be said therefore that race is not important?

Notice the population screened for sickle cell. States are legislating to screen blood populations for sickle cells. It is true that people who are categorized as white have it sometimes. But cost-effectiveness is operating here, and so we have this peculiar conflation. We are talking about the molecular level of risk populations and the interpretation, which is pre-interpreted, of a category called race. One can see why there might be some confusion here. On the posters at this conference, one can see the same conundrum about race and the location of certain haplotypes.

What this has done is to make it very difficult for people to sort out the meaning of race in science. Five years ago, Michael Klag (1991) and his colleagues reported on hypertension in the black community. They found out that, in general, the darker the skin color among blacks, the greater the level of hypertension. There were some exceptions, but that was the general finding. The study concluded that the issue was not about race but rather about the tension that people who are darker experience in America. So here is a published study of hypertension in which the investigator had to disclaim the role of biology in order to get at the interpretive framework of what it means to be of a darker color in America. Why was Klag so concerned? Well, for a very good reason. I will show you some data which I think will crystallize the point as to why African Americans just stay very alert to this slippery interchange between the molecular and the behavioral, between the phenotype and the genotype, and how scientists themselves sometimes slip back and forth, not very carefully, in a very fuzzy way of not thinking about first and second order constructs.

Here is the way many people theorize about race and behavior. They will see some pattern and take an orderly existing phenotype demonstrated by many individuals and then theorize back to the genes. For experience with this concept, let us take the example of the appearance of criminal statistics and how in recent years there has been a dramatic increase in the number of black people in prisons. Why should this be of concern? Because historically the way one theorized about the genetics of a phenomenon was to take those who were incarcerated and then assume that the gene explained their appearance in these "high" rates. Let me show you why this is a deep issue for African-Americans.

We know from the best research available that for certain crimes there is a fall away rate of about 80%, that is, the criminals are not caught. With car theft, this rate is about 97%; that is, the police just cannot catch people or refuse to catch people who are car thieves. However, for certain kinds of crimes there is a huge fall away rate; with burglary it is about 80%.

In Skolnick's 1966 study of a police department in a major city in the U.S., less than 25% of the burglaries were cleared by arrest and prosecution, much less conviction and incarceration. The figure was only slightly higher for robberies, a repeated pattern in reports from around the nation (President's Commission on Law Enforcement, 1967), and this has not changed in the last two decades (Federal Uniform Crime Report, 1988).

To understand the sieve of the criminal justice system that produces the remarkable skew of human subjects that show up in prisons, it is necessary to move from the commission of the act characterized in law as a "crime" to the point of conviction. Starting with the arbitrary figure of 1,000 burglaries, it is generous beyond the best empirical research now available to say that 700 will come to the attention of the police (Reiman, 1984). Of these, at most 300 will be "cleared by arrest," and a maximum of 180 would go to trial. Of these, at most, 120 would be convicted (often, plea bargaining lessens the "crime" to a different category). Of these, no more than 75 will ever spend any time in prison. To designate this the "criminal population" for purposes of research is obviously not a sound procedure. Yet, it is from these records that researchers come along to obtain their data on the "genetics" or "biology" of criminals.

Figure 1 represents my home state of California, the Office of the Attorney General's Report to the Department of Justice, a profile of arrests and incarceration. On the left, you see that African Americans are about 7% of the population in California; yet, we are 18% percent of those arrested and 32% of those sent to prison. Whites are about 54% of the population, 35% of those arrested, and 29% of those imprisoned. That gives you a picture of what I call the second order construction of race. It is a salient category that can be studied empirically and indeed therefore scientifically.

Figure 2 is also from California showing by race the percentage of people aged 20 to 29 in the criminal justice system: African-Americans 39%, Whites 5%, and Latinos 11%. What is the salience of this study to human genetics? Well, when my colleagues tried to call the study "the phenomenon of violence and imprisonment" and connect the fact to race, they went to populations that were at hand, that is, incarcerated. They did not go to the general population but to the population that is known to be criminal, and then they theorized back from that population to what they think are the processes that will explain it.

The controversy of the extra Y chromosome is based upon a population that was incarcerated back in Scotland in 1965. Out of that came some behavior

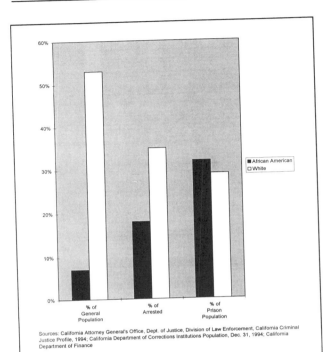

Fig. 1. Arrests vs. Incarcerations, by Race. Sources: California Attorney General's Office, Dept. of Justice, Div. of Law Enforcement, California Criminal Justice Profile, 1994; California Dept. of Corrections Institutions Population, Dec. 31, 1994; California Dept. of Finance.

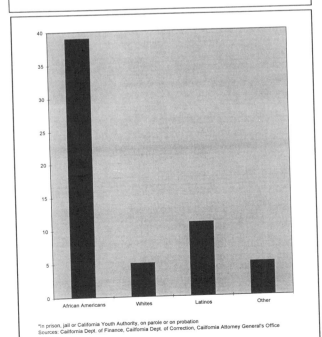

Fig. 2. Percentage of California men ages 20-29 in the criminal justice system (In prison, jail or California Youth Authority, on parole or on probation), by race or ethnicity. Sources: California Departments of Finance and Corrections, California Attorney General's Office.

that women in 1993 were still using—elective abortion—when they found out that their fetus had the extra Y chromosome. In other words, there are consequences in this second order construction to the concept that I think some scientists are regarding as first order.

What is happening with race is that we are confusing these two levels. One can do empirical, scientific studies of the implications of race at the second order of construction, but not at the first order. But that leads people to make a confusing leap. They get confused about how race can be the subject of empirical investigation if it is epiphenomenal, a master narrative, not neurophysiological or biochemical. My point again is that it depends upon the construction level of the concept. What I am suggesting is that the field of behavioral genetics has made remarkably good use of this confusion. It has begun to take a look at behavior and theorize back to the gene. When it comes to African Americans and issues around performance on intelligence tests, appearance in prisons, proportionate frequency on welfare rolls relative to the population, one can begin to see how people are confused about this matter.

I close with the summaries of many of my colleagues who spoke at this conference. The answer is not to stop the machine; it will not be stopped. Millions of dollars are being spent on the Human Genome Project. That means that there are careers which are being pursued and pushed, that means that people have lots of staked interest in this project. These are not necessarily bad things, but this train is moving. Our task is not to be fooled into thinking that this is neutral territory; this is dangerous stuff. It is about behavioral-molecular conflation; it is about people saying that if sickle-cell anemia is in the black population and that is genetic, then why not crime? There is the confusion of first and second order constructs.

My students at Berkeley come to see me and say, "Professor Duster, you are in this area, please tell me. I have a roommate, a cousin, a friend who believes about basketball players in the NBA, that the fast-twitch muscle is genetic. What do you think?" These are students at Berkeley in social science who have come to believe that, when they see these stories about a fast-twitch muscle, there is something to it.

There is one last story. It appeared in the *San Francisco Chronicle* (September 14, 1994) and was headlined: "Bannister—Blacks Have Sprinting Edge." Roger Bannister, who in 1954 became the first person to break the four minute mile, said he was willing to risk political incorrectness by claiming that black sprinters have a certain natural anatomical advantage over their white rivals. Bannister said that it was possible that black athletes' muscles were better adapted to hot climates, etc. He went on to suggest that they could have better power-to-weight ratio because they have less fatty tissue under the skin. Bannister, a 56-year old retired neurologist, gave no evidence to support this position. Bannister was a speaker at a conference of the British Association for Ethical Science when he made these comments on the vast number of black sprint finalists and a high proportion of other black athletes winning world championships.

Here is what Bannister says: "Perhaps there are anatomical advantages to the length of the Achilles' tendon, the longest tendon in the body. I do not know the true reasons. As a scientist rather than a sociologist, I am prepared to risk political correctness by drawing attention to the seemingly obvious but under-stressed fact that black sprinters and black athletes in general all seem to have certain natural anatomical advantages." So here is Bannister claiming the mantle of science that he understands. If incorrect politically, he can make the conclusion that sociologists do not engage in science, but that he does.

## References

Kaufman, Felix. 1958. Methodology of the Social Sciences. New York: Humanities Press.

Klag, Michael, P. K. Whelton, J. Coresh, C. E. Grim and L. H. Kuller. 1991. The association of skin color with blood pressure in U.S. blacks with low socioeconomic status. J. Amer. Med. Assn. 265(5):599-602.

Reiman, Jeffrey H. 1984. The Rich Get Richer and the Poor Get Prison. New York: John Wiley & Sons.

Skolnick, Jerome. 1966. Justice Without Trial: Law Enforcement in a Democratic Society. New York: John Wiley & Sons.

# 5

# Education and the Human Genome Project

# THE VIRTUAL WORKPLACE: A MODEL BIOTECHNOLOGY CLASSROOM PROGRAM

**Robert Yuan and Spencer Benson**

## Introduction

As scientific research and its spinoff in high technology industries continue to grow rapidly, there has been increasing concern about the adequacy of the U.S. educational system from primary school through graduate school. Our university experience in the biological sciences and biotechnology has enabled us to define certain principal issues:

- the need for a more effective and practical means to teach modern biology to undergraduates who are nonscience majors;
- the need for developing an interdisciplinary approach in courses for scientists and engineers in general and biologists in particular; and
- an emphasis on teaching an ever-expanding body of facts at the expense of requiring higher order analytical and synthesis skills.

## Objectives of This Concept Paper[1]

Education at the undergraduate and graduate levels continues to operate on the assumption that careers and job markets in the biological sciences have not changed fundamentally in the past two decades. However, current graduates go forth into a working environment characterized by:

- interdisciplinary approaches to projects;
- working in teams;
- rapid turnover in projects;
- rapid growth of scientific information;
- layoffs and need for retraining;
- expanded career options; and
- the international nature of science and technology.

The authors are professors in the Department of Microbiology at the University of Maryland, College Park, MD 20742-4451. This concept paper was disseminated and implemented in September, 1995. Portions of this paper are being published in one of the essays in the book entitled *Essays in Quality Teaching*, James Greenberg and Stephen Selden, ed.

Universities have moved very slowly in preparing their graduates for an increasingly complex, challenging, and changing world. It is our purpose, in this paper, to present a series of concepts and initiatives that address these issues. The bridging concept is *the virtual workplace*, an undergraduate and graduate curriculum that would provide students with a spectrum of scientific thought processes and skills that prepare them for a variety of scientific careers. The other concept that we have begun to explore is called *journey without maps* that relates to the increasing diversity of both the student body and faculty. For many minority students, acquiring an education in an unfamiliar and complex university system, is indeed a journey without maps. A successful approach to the challenge of diversity would enable the students to collaborate closely toward common goals while accepting each other's differences. These ideas have been tested, in part, in the biological sciences curriculum at the University of Maryland at College Park. The establishment of new courses and the modification of existing ones appear to have been successful as measured by the performance of the students, the acquisition of targeted skills, and by student evaluations of the courses.

Our main objectives in these biology courses are to teach students how to:
- acquire and use information;
- integrate information and knowledge from different disciplines;
- learn and apply necessary career skills, e.g., writing and oral presentations;
- work effectively in teams;
- participate in and be subjected to peer review; and
- be flexible in the use of science for different objectives.

## An Approach to Undergraduate Education

To integrate the concepts of the *virtual workplace* and *journey without maps* into our biology curriculum, we have defined three principal areas for educational innovation: content/process, skills, and work environment. In content/process, we have attempted to cover the principal scientific concepts for each topic while emphasizing problem solving (both in assignments and examinations) and use of literature (including original papers) other than textbooks. The skills that have been incorporated are: working in groups, data analysis, project design, writing, oral presentation, and peer review. We have attempted to approximate the work environment by the extensive use of electronic communications, increasing the sensitivity to diversity, empowering the students to make certain decisions in the course, and making the material relevant to their professional goals and their personal objectives. It has been our intention to develop a set of courses as test vehicles into which we have incorporated these concepts.

It is important to draw a distinction between a major set of courses for science and biology students, and a specific course for nonscience majors. We shall begin by briefly focusing on the course for nonscience majors. This course is described in a recent ASM News publication.[2]

## A Biology Course for Non-Science Majors

*MICB100, Microbes and Society,* is a lecture/laboratory course (39-130 students) in modern biology for nonscience majors, most of whom are first year students. This revised course has now been taught for six semesters. The course is designed around a set of modules that are relevant to the students in terms of everyday life and personal experience. Students are required to read reviews and articles from nonspecialized literature. The tests focus on the use of information rather than memorization promoting a better understanding of the scientific process.

Teamwork involves organizing the class into groups of four students that work on problems and quizzes. Writing takes the form of open book essay exams, laboratory questions and an essay that is half of the final examination. This essay is a take-home exercise that students may discuss with others. The essay asks the student to indicate his/her major/career interests, long-term life objectives, personal interests. He/she then selects a topic covered in the course and dedicates one or two paragraphs to describing the topic. This is necessary so that both student and instructor have a clear idea of the topic being presented. Finally, the student is asked to describe how this topic is relevant to him/her personally. In order to do well in the essay, the student has to show an understanding of the topic and integrate it into another discipline or some specific incident in his/her personal life.

The laboratories in MICB100 are designed for nonscience majors. They have several major constraints:

- time limitations—there are only two one-hour sessions per week;
- the need to use only simple equipment, and reagents; and
- the students' lack of previous laboratory experience.

Each of the laboratory experiments is intended to provide a good sense of how researchers would approach a particular type of problem, and the type of results that would be generated. The concepts illustrated by each experiment are linked to the lecture modules. In several exercises, students are given minimal information and asked to design simple experiments to test hypotheses. This type of active learning reinforces the concept that science is discovery, not memorization of facts.

## Biology Courses for Biological Sciences Majors

Each of these courses fulfills a specific function in the curriculum:

*BIOL105, Principles of Biology I,* is a very large (>1000 students per semester) introductory lecture/laboratory biology course for biological sciences majors and majors in other disciplines, e.g., chemistry, psychology, and health sciences. There are >400 students per lecture section. Lecture sections are taught by faculty members from several departments. The BIOL105 teaching team (faculty members who lecture in the Fall and Spring semesters plus the laboratory coor-

dinator) meet regularly to oversee the textbook selection, course content, sylla-
bus format, test structure, course rules, and laboratory experiments. The exact
content and order of the lecture topics and test format are determined by each
instructor. This team approach ensures that there is 80 to 90% uniformity among
the lecture sections and within the course from semester to semester while allowing for the personalization of content by each instructor. The laboratory is
coordinated by a full-time staff person and has 40 to 50 sections per semester
with 20 students per laboratory. The laboratories are taught by graduate teaching assistants, TAs.

*MICB200, General Microbiology,* is a large (>600 students per year) introductory lecture/laboratory microbiology course for biological sciences majors and
students in other science disciplines. There are 150 to 350 students per lecture
section. Lecture sections are taught by several different instructors. The laboratory is routinely coordinated by one of the instructors. There are 15 to 25 laboratory sections with 18 students per section each semester. Laboratories are
taught by TAs. BIOL105 is a prerequisite for this course. The pedagogy used in
MICB200 includes: student projects, teamwork, peer review, written exercises,
and the use of scientific examples that exemplify diversity.

*MICB380, Bacterial Genetics,* is a medium size (35 to 40 students) upper level
lecture/laboratory course on procaryotic genetics. It is designed as an intensive
bacterial genetics laboratory course. The prototype was run with 12 students as
an inquiry-based course using a combination of classical and current research
problems as the teaching platform. The lab concludes with a student-designed
laboratory in which each student team defines an experimental problem and
designs a set of protocols and experiments to address this problem. This laboratory course is being expanded to its normal capacity of 40 students.

*MICB453, Recombinant DNA Laboratory*[3], is a medium size (30 to 40 students) upper level laboratory course that focuses on basic recombinant DNA
technology. This class uses laboratory driven exercises to demonstrate various
recombinant DNA techniques. Discussions focus on data interpretation. Exams
are essay type and involve problem solving questions designed to integrate the
various techniques demonstrated in the laboratory experiments.

*MICB460, General Virology*[4], is a large (50 to 100 students) upper level survey
course on animal virology. MICB460 has been restructured to use instructor-
assigned student teams that work on a research project and do peer review. This
course is in the process of introducing additional group projects that focus on
data analysis and original research articles.

*MICB470, Microbial Physiology,* is a medium size (25 to 40 students) upper
level lecture course on bacterial physiology. It has been redesigned to use mixed

**Table 1. Task for Skills Development: Undergraduate Courses**

| | BIOL105 | MICB200 | MICB380 | MICB453 | MICB460 | MIC470 |
|---|---|---|---|---|---|---|
| Supplemental Readings[a] | Popular press[b] | Periodicals | Original papers[c] | Original papers[c] | Original papers[c] | Scientific reviews |
| Writing Assignments | Short essay[d] | Short essays[T] Personal final essay[e] | Lab reports Project summary[T] | Short essays Lab reports | Primary literature paper[T] Data analysis[T] | Research proposals[T] |
| Oral Presentations | | | Laboratory progress reports | | | Project presentations and reviews |
| Peer Review | Yes | Yes | Yes[d] | Yes | Yes | |

[a]Assigned reading in addition to the textbook.
[b]Articles in newspapers, magazines and books for the general public.
[c]Articles in scientific journals, e.g., *Science, Journal of Bacteriology*, etc.
[d]Essays of 1-3 pages in length.
[e]An essay that relates the course material personally to the student.
[T]Indicates that the task is done as a team effort.

student teams. These teams prepare and write project proposals that are presented orally to the whole class and are subjected to peer review.

In the upper level courses most of the students are seniors in the biological sciences and biochemistry. These courses are also taken by graduate students.

These courses have undergone revisions in recent years. The revised courses have been taught for the following lengths of time; BIOL105, one semester, MICB200, one semester, MICB380, two semesters, MICB460, one semester, MICB470, two semesters. MICB453 was introduced as a new course offering in 1985. It has been taught for seven semesters.

Table 1 summarizes the features that have been incorporated into each course. The size and level of the course determines which features have been used.

Pedagogy

*Content/Process*

Courses are designed as a set of modules to allow for changes in individual lectures without interfering with the overall flow of the course. The basic premise is that scientific information will accumulate and change at a fast rate, and that university graduates must learn how to selectively acquire new information and use it to solve problems. In all of the courses, students are responsible for critical reading of text and nontext material and

**Table 2. Features of Biology Undergraduate Courses.**

| | BIOL 105 | MICB200 | MICB380 | MICB453 | MICB460 | MICB470 |
|---|---|---|---|---|---|---|
| Problem Solving | + | + | +++ | +++ | ++ | ++ |
| Primary Literatrure | | | ++ | + | +++ | +++ |
| Teamwork | + | ++ | +++ | +++ | ++ | +++ |
| Data Analysis | | + | +++ | +++ | ++ | +++ |
| Project Design | | | +++ | | | +++ |
| Writing | + | ++ | +++ | +++ | ++ | +++ |
| Oral Presentations | | | ++ | | | +++ |
| Peer Review | | + | ++ | | + | + |
| E-communications | + | + | + | | + | + |
| Diversity | | + | ++ | | ++ | ++ |
| Empowerment | | + | +++ | | + | ++ |
| Relevance | + | + | ++ | ++ | ++ | ++ |

Scale   + used in one task or assignment
++ used in two tasks or assignments
+++ used in more than two tasks and/or at a higher degree of complexity

careful note-taking. Assignments and tests focus on the use of information rather than memorization. In all of the courses, we use some tests that combine multiple choice questions and short essays.

The critical reading of non-textbook material ranges from articles in the popular press (BIOL105) and general review articles (MICB200) to technical mini-reviews and original research papers (MICB380, MICB453, MICB460 and MICB470).

*Skills*

In our experience, students find courses to be more relevant if in them they master skills (see below) that are important in daily life. The courses are structured to train students in these skills and to require increasing levels of proficiency in these skills (Tables 1 and 2).

*Teamwork.* In BIOL105, groups of four students work in class on pop problems and quizzes. In MICB200, the students are assigned to lab groups of four students that also work on lecture problems and essays. In MICB380, the instructor-assigned lab research groups (three students) work together on laboratory experiments, problem assignments, and on the design and execution of a student formulated laboratory experiment, which includes written and oral progress reports. In MICB453, students work together on laboratory projects. In MICB460, student teams work on a group project that leads to a research paper. In MICB470, a four-person team is the basic study unit that reads and discusses the reviews and prepares research proposals.

*Data Analysis.* This critical activity is incorporated into the problem assignments and tests given in all of the upper level courses. In addition, it is a fundamental aspect of the laboratory component of MICB380 and MICB453.

*Project Design.* In MICB380 and MICB470, students develop a research proposal that includes defining an objective(s) and an experimental approach as well as ways to interpret the data that is generated. For MICB380, this is done in the context of actual experiments (both protocols for regular experiments and the student-designed experiment). For MICB453, students must evaluate data and determine plausible reasons for the variation seen in the results obtained by different laboratory pairs.

*Writing.* In BIOL105, students have to write a laboratory report in standard publication format and critique a science news article. The class size (>400 per section) precludes additional writing assignments. In MICB200, each group works on a problem set, two short scientific essays and, as part of the final, a major essay similar to the one used in MICB100 (see above). In MICB380, each student writes two individual lab reports based on group data. In the first report, a rewrite is allowed following a review by the instructor. The group writes a progress report at the conclusion of the student-designed laboratory project. In MICB453, students write laboratory reports and work on laboratory-based essay questions. In MICB460, each group works on a project that results in a research paper. Following peer review, the paper can be rewritten before being submitted in its final form. In MICB470, each group writes four research proposals based on mini-reviews. The four page proposals include background, objectives, experimental design, conclusion. While three of the proposals are routinely based on scientific reviews, the last one has involved an evaluation of a business plan by a new biotechnology company.

*Oral Presentations.* In MICB380, students provide overviews of the experiments and progress reports on the final student-designed laboratory projects. In MICB470, members of each group take turns presenting their research proposal to the class or, alternatively, they do an oral review of another group's presentation. These are full-dress presentations involving the use of overhead projections, time limits, and questions from the audience and the instructor.

*Peer Review.* An important aspect of scientific work involves the critical review of the work of colleagues, as well as accepting constructive criticism as the basis for improving one's work. In MICB470, oral presentations and research proposals are subjected to peer review. Each proposal is given a group grade. Each member of the group evaluates the other group members according to effort and productivity. These peer evaluations are then used to convert the group grade to individual grades. A similar system is used in MICB380 and MICB460.

*Work Environment*

It is important that students familiarize themselves with certain characteristics of the modern work environment. Their eventual productivity in the workplace will depend in part on how comfortable they feel in that environment.

*Electronic communications.* The use of computers and electronic telecommunications (internet) is now a fundamental aspect of all scientific work. Electronic listeners (electronic class bulletin boards) and E-mail are of special im-

portance in the teaching of large enrollment courses such as BIOL105, MICB200 and MICB460. Students in the above courses are encouraged/required to subscribe to class listservers. These listservers are used for:
- student-student and student-instructor communications;
- class announcements;
- distribution of lecture summaries, study questions and reviews; and
- dissemination of general interest information, i.e., seminar announcements, recent news articles, interesting internet sites.

Even at this relatively simple level, this type of electronic communications increases the extent of faculty-student interactions at a time when demands on faculty time are on the increase and provide a new informal forum for students to become better acquainted with their instructor and each other.

*Diversity.* This is one of the greatest challenges faced by universities. The student body diversity includes ethnicity, race, culture, gender and academic ability. This diverse student body has to be educated and learn to work together successfully toward a common goal while still recognizing their individual differences. We have explored certain mechanisms in approaching this challenge. Illustrative examples are:
- In MICB200, we are developing a more heterogeneous set of examples for basic scientific concepts (e.g., the making of soy sauce rather than yogurt as an example of fermentation; preparation of sashimi as an example of bacterial cross contamination).
- In smaller enrollment courses such as MICB380 and MICB470, students are asked to write brief biographies and post them on the class listserver so that their fellow students can become acquainted with their backgrounds.
- In MICB380, MICB460 and MICB470, student-teams are instructor-selected so they have a mixed composition as defined by ethnicity/race, GPA, and background in selected biology courses.
- In all of the courses, some personal mentoring from the instructors is often required for students from minority groups who have difficulties with English and written and oral language skills.

Because of the instructor-imposed mixed composition of the student teams, diversity is an integral part of teamwork and peer review processes. By working in mixed teams, students often begin to recognize commonalities that cross diversity lines and that each individual brings to the group a unique set of experiences and skills.

*Relevance.* A key aspect of the described courses is making the course material relevant to the students in terms of the real-life environment to which they will be exposed. Students are more likely to be motivated to learn if they perceive the material to be relevant to both their career and personal goals and can be identified with personal or family situations.

The type of relevance varies depending on the course. In BIOL105, relevance is illustrated by stories from the daily press, weekly magazines and TV news. In MICB200, relevance is stressed with respect to scientific discoveries and their impact on career opportunities and modern quality of life. In MICB380, relevance is illustrated by problems presented in research papers and the development of laboratory projects. In MICB453, the fact that the laboratory exercises mimic those that are routinely done in the biotechnology marketplace is stressed. In MICB460, the association of genetics and replication of viruses to human disease is clearly presented. In MICB470, the relevance of new research areas and their implications for career opportunities is stressed.

*Empowerment.* Students are also motivated when they sense that they have a measure of control over what they are learning. In its most dramatic form in MICB380, students can formulate their protocols and ultimately design and execute their own final research project. A similar freedom of choice is associated with the research proposals in MICB470. At a simpler level, students have also been allowed to choose the topics of the last lectures in MICB200, MICB380 and MICB470.

*Laboratory Experience*

BIOL105, MICB200, MICB380 and MICB453 are laboratory courses. The laboratories in BIOL105 and MICB200 are limited by: (1) a large number of laboratory sections necessitating the use of many graduate teaching assistants who have a wide spectrum of abilities, commitments to teaching, and training; (2) a student population with a wide range of abilities, motivations, and background knowledge; (3) the cost of laboratory equipment and supplies; and (4) appropriate laboratory space.

Due to enrollment caps, the laboratories in MICB380 and MICB453 do not suffer from size-related problems. They are limited by time constraints, cost of equipment and materials, and the ability to find qualified teaching assistants. Our approach to some of these problems has focused on graduate training programs and the extensive use of student team work.

**Evaluation**

Our goals in designing this set of biology courses are to train these students in critical thinking and scientific skills and expose them to the environment of the workplace. Given the complexity of this integrated approach, it is of utmost importance to evaluate its success. Table 3 summarizes the guidelines used in our evaluations.

Overall grade performances were not lower than for the courses prior to their restructuring— indicating that the students are able to meet the demands made on them. In some courses, the class averages are actually higher. The separate grading of course assignments, e.g., research proposals, laboratory projects, and oral presentations allows us to monitor the acquisition of specific skills. In those tasks that are repeated, it is possible to see progress being made during the semester. It is important to note that in the case of MICB470, three of

233

**Table 3. Course evaluation guidelines.**

| Mechanism | Parameter Measured |
|---|---|
| Final course grade | Overall performance in the course |
| Scores on individual tasks | Student performance in different skills |
| Peer evaluation | Teamwork<br>Diversity |
| Course evaluation | Provides student reactions to:<br>- instructor<br>- subject content<br>- assignments<br>- oral presentation<br>- electronic communications<br>- teamwork |

the research proposals were along standard scientific lines while the fourth one involved the novel task of evaluation of a business plan (with a major scientific component) for a new biotechnology company. The students actually performed better, indicating that they had acquired the ability to transfer learned skills to a different type of project.

Peer evaluation allows the conversion of a group grade to an individual grade based upon contributions to the team. It also provides an invaluable window into the functioning of individual groups. Finally, the special course evaluation form used provides us with:

- students' overall view of the course;
- an estimate of effort required for the various tasks;
- the perceived value of each task;
- an individual view of progress made; and
- criticisms useful for further course modifications.

Overall, the students thought that the courses were difficult and forced them to attempt new tasks to which they had never been exposed. Nevertheless, students indicated a sense of deep satisfaction in being able to carry them out satisfactorily and in the context of their own team and the larger group.

Lastly, personal comments from two students give insights into how individuals cope with diversity:

*A student in MICB380*: I remember going to you to discuss about my group's ability to communicate well with each other. It was the second day of the lab and you told me that I shouldn't worry.

You were right. Although I learned quite a lot about microbiology, I know that the deepest lesson learned here was to really in all proportions work as a *group*. As the world moves toward an international environment, having the other foreign students was *fun*. Somehow make sure that we all understood everything *together*. Yes, Bei needs some more practice in her verbal skills and Oleg needs to be more aggressive but my group's situation made me remember what it was like for me when my English wasn't that good. It made me remember to be patient and understanding. It also made me remember when I wanted to communicate with everybody. This group was a challenge. The other reasons that just came to my mind was that I forgot what it is like to be in Bei's shoes and feel the frustration of not being understood well enough. At times I even corrected her. That felt good!

*A student in MICB470*: Aside from the 'school knowledge,' I learned an important lesson on how to work with groups. I took your advice and first talked to Zeba to find out what her "personal" problem was with me, and then I addressed the group as a whole. Everyone was really nice, and we talked everything out. Now I know that, when I am in a group situation, personal problems have to be resolved right away; otherwise the work gets affected.

I have always worked well in groups and have gotten along with my group members. I have never been in a situation like this before. In a way I'm glad this happened because now I know I can deal with these situations down the line. I'm happy with the way I handled everything. Anyway it was a very tough class but well worth it.

In one case, a foreign student saw the experiences of other team members in the context of her own learning experience. In the other, team members minimized the contributions of an attractive young female student (presumably on the assumption that she was just a pretty face). She was able to work out her problems with her group and in doing so improved the group's performance. These comments provide hope that the use of integrated teams may be an effective approach to the challenge of diversity.

**Conclusions**

As summarized in Tables 1 and 2, the undergraduate courses share certain common features. They have been designed to provide an overview of biological information anchored by a more detailed exposure to specific topics. The students are taught specific skills at increasing levels of difficulty, e.g., finding a solution to a quiz problem → writing essays → writing research proposals, as they move from lower to higher level courses. In the senior level courses, the assignments approximate the activities that they will encounter in graduate school or in work situations (experimental design, group interactions, oral presentations, peer reviews). Throughout this series of courses, a major effort is made to make the content and the skills as relevant to real-life experience as possible while providing an exposure to many of the characteristics of a working environment.

The fact that our approach has now been used in a set of six different courses with apparently positive results is a matter of some satisfaction. However, this is tempered by the recognition that these courses require an increased level of faculty time compared to the more standard lecture-test format. The increased efforts required of the faculty raises the question of sustainability of these educational initiatives. This is particularly true when we move some of the tasks from smaller to larger courses or when enrollment increases due to student demand for our upper level courses. It is our belief that a solution to these problems lies in improved use of electronic media and in a serious effort to train graduate assistants for service in these courses.

While we have endeavored to create a series of integrated courses that move up a gradient of content and complexity, we believe that single or multiple features can be incorporated into individual courses at any institution. This should enable us to disseminate what we have learned in our teaching experiments to colleagues throughout the United States, as well as around the world.

## An Approach to Graduate Education

An Overview

Ph.D. programs in the biological sciences often have the goal of cloning the mentors, i.e., preparing students for a career in academic research. However, several studies show that approximately half of the Ph.D.'s in the sciences will be working in nonlaboratory occupations 10 years after they get their degree. Furthermore, the existing mindset elevates basic research while ignoring potential careers in teaching and industry.

A different approach to the Ph.D. degree would be to draw an analogy with the automobile industry. When an automobile company decides to develop a new car line from the beginning, it must first create a platform (defined as the basic structure and chassis). The various models in this new line are then built around this platform by combining various components (i.e., engine, suspension, body, brake systems) to give the required performance characteristics (e.g., sport handling vs. comfortable ride). We propose a similar approach to the Ph.D. degree. The platform in this case involves a set of required graduate courses, the candidacy examination, and a thesis based on original research on a specific topic. This platform can then provide three separate options: basic research, teaching, and management.

At present, there is only one option: basic research. The creation of the two other alternative career tracks would require the establishment of appropriate graduate courses, internships, and either the design of an integrated thesis project or an additional project focusing on teaching or management. It must be emphasized that selecting one career track does not exclude training and exposure to the other ones any more than having a station wagon precludes good road-holding characteristics.

The most efficient and intellectually elegant approach to establishing a novel Ph.D. program would be to create an interdisciplinary program housed in an academic department. The realities of academia make this a daunting task in the best of times. We and colleagues in the Technology Management Program of University College have come up with an alternative approach.

The Research Track

The principal focus of most Ph.D programs, ours included, is on research. Our efforts have been principally directed toward modifying the existing approach by emphasizing critical thinking, experimental design, and the same type of key skills that we have introduced into some of our undergraduate courses. An additional novel component would be the creation of a fundamental interdisciplinary capstone course in biotechnology. This fundamental course will combine information from basic science, technology development, commercialization/applications with the case study approach. The course will be structured in six modules as diagrammed below.

Biotechnology: The science      Case Study #3
Case Study #1      Case Study #4
Case Study #2      Societal impact

The first and last modules will anchor the course while the other four modules consist of case studies (see next section). The first module will review the fundamental scientific principles underpinning biotechnology. The goal will be to present general principles and fundamental concepts using original papers. The last module will present an overview of the commercial, economic, legal, political, and ethical effects of this new technology. The course will also emphasize that the impact of this new technology is not limited to the United States, but is global in scope. Two new graduate courses (MICB688N and MICB688P) have provided us with valuable experience related to the development of this new course.

Case Studies

The case study approach has been enshrined in the teaching of law, business, and medicine. We intend to use it for the interdisciplinary teaching of biology. Each case study will focus on a specific project area (e.g., the production of genetically-engineered insulin). Unlike the more traditional use of the case study, the content will cut across different disciplines. We envision each case study to have five components:

1. The basic research that led to the development of a particular product or process.

2. The development of alternative systems (e.g., cloning a gene into

mammalian cells following its successful expression in bacteria) and scale-up for production.

3. The factors involved in the commercialization/application of the particular product/ process.

4. Bioethical and political issues affecting acceptance/rejection of products/processes.

5. Regulatory, legal, and economic impact of the product/process.

Components (2), (3), (4) and (5) will emphasize the process of evaluation, analysis, and decision making.

It is our intention to establish a set of 12 detailed case studies. While at the beginning these would be used in the graduate course, we intend to explore their use in some of the undergraduate courses.

## MICB688N Biotechnology: Theory and Applications

This advanced graduate course was established and taught for three years. The course was designed around four modules: genetic engineering, monoclonal antibodies, bioprocessing, and biosensors. The overall sense of the course was conveyed by lectures given by the instructor interspersed with presentations by visiting lecturers, each of which was an expert in a specific field (e.g., cloning of interleukins; design of biosensors). Outside speakers gave a lecture(s) and had an informal luncheon with the students.

MICB688N takes the training in key skills to a higher level (see Table 1). Students were not only expected to follow the lectures carefully, but were also assigned papers relating to each module. Writing skills were honed by having each student write a review of the paper following the guidelines to the referees of Proc. Natl. Acad. Sci. (PNAS) manuscripts. Some of the papers were deliberately chosen because they would *not* meet the standards of PNAS. Negative reviews were then used so that students could learn how to respond to referees' criticisms. Reviews of the papers were used for oral presentations. No traditional examinations were given in the course. A feasibility study was required for the midterm, and writing a full fledged NIH grant proposal was the final. Both tasks were group projects allowing for both sharing of the work and exposure to teamwork. The NIH proposal was chosen by each team but had to focus on one of the topics covered in the course. The proposal included sections relating to staffing, CV's, and budgets. Oral presentations of the feasibility studies and the grant proposal provided a basis for peer review. It is expected that our experience with this course will provide the basis for the new graduate course.

## MICB688P Genetic Approaches to Complex Biological Problems

This advanced graduate course is taught using Socratic pedagogy. The course uses historic and current research papers to illustrate experimental approaches instrumental in solving key biological problems, e.g., solving the genetic code. On average, students read two to four papers for each class. One session is devoted to ethics in biological research while the topics of the last two sessions are selected by the class. As a midterm exam, the students review a manuscript following journal guidelines. The final consists of the preparation of a postdoctoral proposal. The class serves as its own grant review panel. Each student is assigned as a primary reviewer for one proposal, and as the secondary reviewer for a second proposal. The primary reviewer must prepare a written critique of the proposal and present it orally to the panel. The secondary reviewer then provides additional comments and a confirming or dissenting opinion. The panel then discusses the proposal (each student is expected to have read all of the proposals). Following this, each student assigns a priority score to the proposal. This score is used in determining the grade on the proposal. The proposal grade makes up 50% of the course grade. The remainder of the grade is from the midterm (20%), class participation (15%) and peer-review (15%). The course stresses skills in reading of research papers, presentation of concepts, reviewing of manuscripts, and grantsmanship.

The Education Track: Some Ideas

It is the established view that competence in research (as evidenced by a traditional Ph.D.) provides an adequate background for teaching. Operationally, this is an unproven assumption that also helps to downgrade teaching as a career track for young scientists. Teaching skills can be broadly defined to go beyond an academic pursuit and to be an important component in research, industry and management.

We are thus faced with a dual challenge. The undergraduate experimental teaching modalities described above are labor intensive and require constant updating and modification (arising in part from interaction with students). Accordingly, if we are correct that these are the trends for the 21st century, we have to expose and train the next generation of teachers to this new way of teaching.

Our work on undergraduate courses shows that many of these techniques (even if labor intensive) can be used with classes up to approximately 100. However, applying them to classes of 200 to 300 will require experimenting with computer-aided instruction as well as the use of graduate students as discussion leaders and evaluators of essays and oral presentations. To do so requires that the graduate teaching assistants themselves be trained as teachers/instructors. It is our view that the education track of the Ph.D. program would provide extensive opportunities along these lines. Thus, we could inspire innovation and motivation in the next generation of teachers and at the same time bring some of the advantages of small group instruction to large undergraduate classes.

A first step in this direction was the use of a required graduate student seminar (MICB788). For one semester, 13 students were involved in three tasks:
- use of textbook material for the preparation of class lectures;
- development of both multiple choice and essay questions; and
- grading (and evaluation) of test questions and essays.

Since many of the students were also teaching assistants in MICB200, the large enrollment General Microbiology course, part of the material was taken from that course in order to assist them in their actual jobs. The time limitations of the seminar (one hr/week) imposed restrictions on the tasks that could be assigned. Nevertheless, they were able to carry out:

- teamwork: the class was divided into groups of three or four students that worked together in the assignments,
- oral presentations: each student had to give one lecture or another form of oral presentation (tests and grading of essays), and
- peer review: each assignment was subject to peer review mainly by members of the other groups.

The results were promising in that the students realized that effective teaching requires mastery of both the course content and a series of skills. Even when they considered themselves to be experienced TA's, they ran into serious difficulties giving lectures and evaluating essay questions. The conclusion was that there was a need for a more extensive course that should be required of all graduate students, but it should be in the context of a regular graduate course rather than a seminar. Our experimentation along these lines could lead to additional professional development courses linked to undergraduate teaching.

The Management Track: The Ph.D./M.S. Program

A new joint Ph.D./M.S. program would combine the scientific rigor of a Ph.D. in Molecular and Cell Biology with the skills of an M.S. in Technology Management. The Ph.D./M.S. program has a precedent in the M.D./Ph.D. programs that prepare students for both clinical work and research. The basis for this novel program is to provide training for four types of individuals:

- research scientists who wish to acquire certain skills necessary for running a large laboratory;
- research scientists who will rise to senior administrative positions in universities and institutes;
- managers for the growing biotechnology industry; and
- administrators in certain key government agencies such as the NSF, NIH, FDA, USDA and DOE.

Such a program would combine an existing Molecular and Cell Biology Ph.D. program consisting of four required graduate courses (plus additional electives) together with the M.S. program in Technology Management with a specialized Biotechnology track consisting of six courses. It is proposed that the students would do one integrated thesis to fulfill the requirements of both the Ph.D. and M.S. degrees. In addition, students would be required to take the capstone course in biotechnology described in the Research Track section.

Summary

Conceptually, we are exploring a Ph.D. program that would expose graduate students to three potential tracks: research, teaching, and management. In all cases, our objectives focus on the analytical use of information, and the acquisition of basic skills. The three innovative aspects of the program would involve the creation of a Ph.D./M.S. program that would incorporate the research aspects of a Ph.D. in Molecular Biology with the specific skills of a M.S. in Technology Management, the establishment and use of case studies in biotechnology, and the use of teaching seminars both as a mechanism for training and an attempt to improve the teaching of large undergraduate classes. We have gained extensive experience from the teaching of two graduate courses and have run our first teaching seminar.

Student Feedback: Some Additional Concepts

In both MICB688N and MICB688P, special course evaluation forms were prepared in order to obtain detailed student feedback. In each case, evaluations were favorable though it was felt that these courses were very difficult in terms of the level of effort required. One of the problems identified by students was that some course material was redundant having already been covered in earlier courses.

In those graduate courses (except for students who had already begun thesis research), students stated that it was very difficult for them to develop an experimental design due to lack of laboratory experience. Yet, their records show that they have had laboratory experience in their courses. However, in the minds of the students, running an experiment in a laboratory course is disconnected from the potential use of that technique for proving a theoretical concept.

These two problems need to be addressed. We suggest that this must be done by teaching the students to acquire and use information to develop new approaches to problems. As to the issue of repetitive course content, the answer does not lie in the direction of single exposure but in the depth of the information provided and the complexity of the tasks to be performed. The issue of lab experience does raise the question of how effective wet labs are in teaching students the use of techniques as tools for acquiring specific types of information. In other words, can some of this be taught conceptually?

Again, increased faculty time (alas, something very much in short supply) can provide solutions to these problems. An important alternative, as proposed in the previous section, is the innovative use of graduate teaching assistants as part of their training in education.

**Conclusions**

The basic premise of this paper is that both information in the biological sciences and the career opportunities in those disciplines will continue to change at a rapid rate. An appropriate curriculum both at the undergraduate and graduate levels must fulfill two tasks:

1. Teach the critical acquisition of information and its use in carrying out tasks primarily of a problem-solving nature; and
2. Build skills essential for scientific activities such as teamwork, proposal writing, oral presentations, and peer review.

Underlying this educational premise must be a broader perspective that these graduates will not only work on different project areas during their professional life, but may also have different career goals, e.g., university faculty, laboratory director at NIH, vice president of a biotechnology firm, etc. Their success in mastering change is dependent to a large degree on being able to teach themselves and to learn from their peers. To learn how to do this well requires a perception that the education that they are being given is both relevant to the world of work and to the human society of which they are a part.

## Acknowledgments

A large number of individuals have given generously of their time for the preparation of this concept paper. In particular, the authors would like to express their gratitude to Dr. Bruce Alberts and Dr. Jay Labov of the National Research Council, Dr. Joseph Perpich of the Howard Hughes Foundation, Dr. Herb Levitan of the National Science Foundation, Ms. Amy Chang of the American Society for Microbiology, Ms. Sheila Tobias, and Dr. Werner Arber of the University of Basel for meeting with us and communicating their ideas and criticisms of this manuscript. At the University of Maryland, Dr. James Greenberg of the Center for Teaching Excellence, Dr. Les Glick of University College, and Dr. Maynard Mack of the Honors Program have been a continuous source of encouragement and ideas. Finally, Dr. Sue Gdovin of the Department of Microbiology has been invaluable in helping us shape our ideas and putting them to paper.

## Notes

1. The views and ideas presented in this concept paper reflect those of the authors and do not necessarily represent those of the Microbiology faculty or of the administration of the University of Maryland at College Park
2. Yuan, R. T. and S. A. Benson. 1995. Modern microbiology for nonmajors: Teaching relevance. Am. Soc. for Microbiology News 61:27-30.
3. We thank Dr. Daniel Stein for providing information on the development and structure of this course.
4. We thank Dr. Susan Gdovin for providing information on her modifications to MICB460.

# EDUCATIONAL PROGRAMS ON THE HUMAN GENOME PROJECT SPONSORED BY THE NHGRI

## Paula Gregory

My topic is the education programs at the National Human Genome Research Institute (NHGRI) for which my office is responsible. I would be very remiss if I did not mention other educational efforts on the Human Genome Project funded by both the Department of Energy and NHGRI. Those include projects for health and other professionals, genome center education such as Dr. Rick Myers' group, Dr. Lee Hood's group where the emphasis is on high-school partnerships, high-school curriculum development, teacher education programs, educational web sites and educational radio programs. Having come from being one who got grants for education to being at NHGRI in education, I want to give everybody as much credit as I can because it is a collaborative effort.

What I would like to talk to you about briefly are the three projects I work on: a short course that we have for faculty from minority institutions which I will talk about in greater detail; the Genetic Self, a workshop that we have for family counselors, people who are pastors, social workers and marriage and family therapists; and our high-school DNA sequencing partnership which is modeled after a program started by Dr. Hood and Dr. Maynard Olson at the University of Washington in Seattle. Our office has not targeted primary care physicians because there are already several people who have grants off-campus to work with that group. Our programs try to fill the gaps and cover what is not being done.

## The Genetic Self Workshop

The Genetic Self Workshop was held this year (1996) for the first time. The idea behind this project was to help those who help families deal with genetic information after they leave a medical geneticist's office. What happens to you the rest of your life when you are dealing with a *BRCA1* gene mutation diagno-

The author is Chief of the Genetics Education Office for the National Human Genome Research Institute at the National Institutes of Health, Bethesda, MD 21740.

sis in your family? How do you handle the inter-family problems that will accompany genetic testing and genetic risks?

We identified the professionals who will be helping families deal with this—social workers, pastoral counselors, marriage and family therapists, and psychologists. What do these people know about genetics? Not a lot. What do these people know about the relevance of genetics to what they do? Well, if they are tuned in to the various media, they may know a little more about it. But the full impact of what is happening may not have yet hit home.

NHGRI brought 25 counselors from all over the country to Washington for a three-day intensive workshop to teach them as professionals about basic genetics and the psychosocial impacts of genetic testing. As a result of our first workshop, the American Association for Marriage and Family Therapy is now going to hold a special session just on the Human Genome Project and genetics and what it has to do with marriage and family therapy.

**High School Student Sequencing**

Our high school DNA sequencing partnership is in its second year. We like to think of ourselves as the East Coast branch of the sequencing partnership. We have partnerships with six schools in the Baltimore, Washington, D.C., Montgomery County, Maryland and Fairfax, Virginia area. We have scientists that work with high school students who are sequencing unknown regions of the human genome. At the end of the year the students come to the NIH and present their research in a poster. People like Dr. Varmas, the head of the NIH, Dr. Rick Klausner, and Dr. Francis Collins, come by and see what teens can do as far as sequencing and genomic science goes.

**The Short Course for Faculty at Minority Institutions**

What I am really here to talk about is our short course in genomics. It is now three years old. The course has two main goals: one is to increase minority understanding and participation in genetic research by teaching college faculty. The second was to increase collaboration between faculty who teach at minority universities and NHGRI scientists.

The course is one week in length and combines both lectures and hands-on work. It has national impact as we bring faculty from institutions all over the U.S. with a substantial minority enrollment. This course is designed to foster research collaboration and cultivate minority student interest in genomic research. This does not necessarily mean just science students. We are also interested in non-science majors, so if they read *Glamour* magazine's article (October, 1996) about genetic testing, they can say, "Oh, yeah, when I was in biology class, my professor told me about that."

How does one get in the class? First, send a letter of interest to Jeff Witherly at NHGRI stating why you want to come and what you plan to do with the information when you get back to your college or university. Second, there needs to be a letter of support. We want to know that your university and your department supports this because it is going to make it easier for you to implement this into your curriculum and tell others in your department about it. We also require that the home institution support the travel of the participant, but we support their food and housing while in Bethesda. Finally, you have to submit your curriculum vitae.

The kinds of topics that we cover in the course are very broad in range. We have experts on campus in a variety of areas of genome research, and we try to touch on everything with the idea that everyone is going to know more about at least one of these topics, but not everybody is going to know all of these topics. Basically we cover cloning technologies especially new developments, physical mapping, sequencing, gene therapy developments, discussions on ethics and case studies of various kinds they can then take back and share with their students. We also have someone talk about complex disease research. We also talk about grant-writing—what do you need to get a grant? Where does it go after it gets to NIH? How can you facilitate the process?

In the past two years we have had 20 participants, 12 in the first year and eight in the second year. The participants come from across the country and teach African-Americans, Hispanics and Native Americans in a variety of settings—in medical schools, small universities, liberal arts colleges, and even community colleges. We tailor the course to the participants. Fifty percent of them so far were members of a minority group, primarily African-American or Hispanic. Thirty percent were department chairs, which can be really important for implementing change in the curriculum. The other thing that we try to do in the short course is to make sure that the participants know that other people from their universities can also apply. We would like to build a critical mass of people at a university who are excited about genome research.

In 1995 we asked the participants to self-rate themselves from the first day about how much they learned during the course. As can be expected, the things people said they learned most about were things new to them, but they also learned more about things they thought they knew a lot about. These evaluations have been helpful to us in getting a handle on what things faculty consider important to their needs and are helping us improve the short course.

The setting for the course is very informal. That is the nice part about it. It is a small group. People come and they stay in Washington, and we try to make sure they have a good time and enjoy being in Washington. We always conclude with a crab fest and in this informal way we can find out more about what they thought of the course and how they felt everything went.

I believe the course gets the message out about the cutting edge research which the Human Genome Project represents as well as the latest developments. In addition, we are working on some distance learning type formats. We are putting together a series of lectures that can be uplinked to various networks.

# THE HUMAN GENOME EDUCATION MODEL (HuGEM) PROJECT: EDUCATING CONSUMERS AND HEALTH PROFESSIONALS

E. Virginia Lapham

The Human Genome Education Model (HuGEM) Project is a joint effort of Georgetown University Child Development Center and the Alliance of Genetic Support Groups to develop a collaborative education model for families and individuals with a genetic disorder (consumers) and for health professionals who provide preventive, diagnostic, counseling, and treatment services for persons with genetic conditions (providers) in University Affiliated Programs (UAPs) across the country. The project was carried out from 1993-1997 and focused on the ethical, legal, and psychosocial implications of the Human Genome Project and related genetic research and testing. The project is funded by the National Institutes of Health's National Human Genome Research Institute, Ethical, Legal, and Social Implications Research Branch.

The rationale for focusing on education of consumers is that individuals with or at increased risk for known genetic disorders and their families are the persons most immediately affected by the Human Genome Project. They are among the first asked to participate in research designed to identify new genetic disorders, the first to be offered new treatments such as gene therapy, the first to have to choose how much they want to know about their future health risks, and the first to find out the insurance and work place implications of having their genetic risks known if confidentiality is not maintained.

Health professionals (providers) such as physicians, nurses, social workers, psychologists, speech pathologists, audiologists, occupational therapists, physical therapists, nutritionists, special educators, and others who work with individuals with or at increased risk for genetic disorders are also affected by the Human Genome Project. These health professionals are often the first to give a label to developmental or behavioral symptoms, the first to recommend further evaluations, including genetic testing, the first to influence client attitudes about genetic research, the first to give genetic diagnoses, and the first to

The author is Co-Director of the HuGEM Project at the Georgetown University Child Development Center, 3307 M Street N.W., Suite 401, Washington, D.C. 20007-3935. ©May not be reproduced in any form without the expressed written consent of the author.

provide counseling to affected individuals and families. The health professionals in the more than 70 University Affiliated Programs across the country were selected as a discrete group of health professionals, primarily associated with medical schools, that provide these services.

The overall goal of the HuGEM Project is to derive optimal benefit from the development of the Human Genome Project for consumers and for health professionals. An underlying premise is that collaborative education will enhance the mutual understanding and cooperation needed for the two groups to work together and to make key decisions about if and/or when each will participate in the Human Genome Project.

The first step in developing an education model for consumers and providers was to develop and implement a national survey that would include priorities for education. This paper focuses on the topics rated as high priority by the 332 consumers and 329 providers who responded to the survey. Although a concerted effort was made to include representative samples of all ethnic groups in the United States, only 10% of consumers and 9% of health care providers defined themselves as African-American, Asian-American, Hispanic/Latino or Native American/American Indian. Nevertheless, since this conference is focused on the concerns of these groups, a separate analysis is presented of the priorities for education of these groups.

## Methodology

A total of 332 consumers from 44 states and the District of Columbia participated in telephone surveys that lasted an average of 40 minutes. They were recruited from genetic support groups across the country. Mail responses were received from 329 health professionals from 52 University Affiliated Programs in 41 states, Puerto Rico, and the District of Columbia.

## Characteristics of Respondents

Consumer respondents were primarily female, highly educated, and married—characteristics believed to be typical of genetic support groups. More than half of the respondents (55%) were themselves affected by a genetic disorder. There was an average of 2.1 affected family members per respondent with a range of one to more than 12 affected members. Overall, the consumers represented 101 different genetic disorders. Ten percent of the consumers (32) identified themselves as African-American, Asian-American, American Indian, or Hispanic. The group of 32 consumers represented 17 different genetic disorders (Ehlers Danlos, Leukodystrophy, HHT, Osteogenesis Imperfecta, Neurofibromatosis, Congenital heart defect, VHL, Oryithine Trans. Carb, Hemophilia, Mucolipidosis, Sjojgren's Syndrome, Spinal Muscular Atrophy, Beckwith Weideman Syndrome, Down Syndrome, Williams Syndrome, TAR Syndrome, Wolf Hirshorn Syndrome, Sickle Cell Disease, Maple Syrup Disease).

The 329 health professionals (including 25 respondents who identified themselves as African-American, Asian-American, American Indian, or Hispanic) had a mean of 17 years work experience and were highly educated. All held graduate degrees with 54% having completed doctorates including 14 of the 25 minorities. The health professions represented by the respondents included: Audiology/Speech and Language Pathology, Education, Medicine, Nursing, Nutrition, Occupational Therapy, Physical Therapy, Psychology, and Social Work. Almost all of the health professionals indicated they provided services to persons with genetic disorders but less than a third reported having one or more semester courses in genetics. The health professionals recognized the importance of genetic testing and research and indicated a strong desire for education in genetics.

**Findings on Education Priorities**

From a list of 15 possible topics for education (See Tables 1 and 2), the top priority for both consumers and providers was *"Coping with a new genetic diagnosis in the family."* The selection of this psychosocial topic indicates the importance of the emotional impact of a genetic diagnosis for both those who are diagnosed with a disorder as well as those who give a diagnosis and/or assist individuals to cope with a diagnosis.

The second priority topic for consumers and third for professionals was *"Treatments for genetic disorders including gene therapy."* The gap between the capability of medicine to diagnose a genetic disorder and being able to provide effective treatments is a concern reflected in the choice of this topic. Persons who are affected with a genetic disorder for which there is little or no treatment at present want to keep abreast of any developments that may he helpful to them and/or their family members. An example was a survey respondent, who at 32 said he was the oldest living person with his genetic disorder, was excited about the prospect for treatments coming out of the Human Genome Project. Even though acknowledging that it was probably too late for him, there was still hope in his voice as he said that at least there was hope for others with his disorder.

The third priority topic for consumers and second for providers was *"Who has access to genetic information?"* The concern for consumers is that access to genetic information by health insurers and employers is that they will lose insurance coverage and be unable to pay for medical and hospital costs of treatments for their disease. Providers are concerned for their patients/clients and face ethical dilemmas of what to write on medical records that will he reviewed by insurers.

The fourth priority for both consumers and providers is *"Genetic disorders and health insurance."* The issue of access to genetic information spills over into this category where the primary concern is health insurance. As one respondent commented, health insurance is a matter of life and death for many persons with genetic disorders.

The fifth priority of the consumers was *"How to stay informed about new developments in the Human Genome Project."* This was less of a concern for health professionals who ranked this topic as number 10. Health professionals may feel that they already have resources for staying informed.

As may be seen in Tables 1 and 2, the remaining priorities of the consumers and health professionals were in somewhat different order.

A separate analysis of the minority consumers and health professionals showed the same top five priorities. To get an idea of the priorities of the minority participants, additional suggestions for education made in an open-ended question at the end of the questionnaire are summarized below each table:

**Table 1. Priority topics for education of consumer respondents (N = 332)**

1. Coping with a new genetic diagnosis in the family. (87%)
2. Treatments for genetic disorders including gene therapy. (82%)
3. Who has access to genetic information. (76%)
4. Genetic disorders and health insurance. (73%)
5. How to stay informed about new developments in the HGP. (70%)
5. Health Care Reform and the Human Genome Project. (70%)
7. New genetic information and the legal system. (65%)
8. Advantages/disadvantages of participating in family studies of genetic conditions. (62%)
9. How society may be affected by the Human Genome Project. (61%)
10. Family and Professional partnerships. (57%)
11. Genetic information and employment. (50%)
12. The media and interpreting results of genetic research. (50%)
13. Including children in decisions about their genetic testing. (49%)
14. Genetic testing and biotechnology companies. (45%)
15. Screening of the general public for genetic disorders. (29%)

Other topics suggested by consumers who identified as a cultural minority:

- Who has access to quality control? What is the federal government doing to curtail dissemination of information?
- Research priorities: How are they determined; Do they lean toward one group: Is there discrimination based on race?
- Computerized information available.
- How to advocate for funding.
- How to get disability benefits with genetic conditions (be successful).
- How consumers can educate physicians on disorders.
- How to funnel information to smaller populated areas and defined neighborhoods.
- Learning how to cope. (Many families break up because they cannot handle.)
- Educating the children.
- How genetic testing can be sensitive to religious beliefs.
- Basic knowledge of genes and how they work. (Average person does not know genetics.) Important to target ethnic/cultural groups.
- How to raise funds for genetic research.
- How professionals can be more considerate and informed about the diagnosis of a child and feelings of parents. How to get HMOs to be more concerned.
- Helping siblings understand and cope with a genetic problem. Education for fathers about genetic conditions.

Table 2. Priority topics for education of health professional respondents (N = 329)

1.  (F) Coping with a new genetic diagnosis in the family. (90%)
2.  (B) Who has access to genetic information. (86%)
3.  (D) Treatments for genetic disorders including gene therapy. (75%)
4.  (G) Genetic disorders and health insurance. (71%)
5.  (J) New genetic information and the legal system. (69%)
6.  (O) Family and professional partnerships. (64%)
7.  (A) Advantages and disadvantages of participating in family studies of genetic conditions. (61%)
8.  (H) Genetic information and employment. (57%)
9.  (E) Including children in decisions about their genetic testing. (56%)
10. (K) How to stay informed about new developments in the HGP. (52%)
10. (M) How society may be affected by the Human Genome Project. (52%)
10. (N) Health care reform and the Human Genome Project. (52%)
13. (C) Screening of the general public for genetic disorders. (43%)
14. (L) The media and interpreting results of genetic research. (38%)
15. (I) Genetic testing and biotechnology companies. (35%)

Other topics suggested by health professionals who identified as a cultural minority included:

- Psychological support, counseling.
- Ethics. How genetic testing was utilized in the past.
- Ethics and cultural/racial implications.
- Strategies for person-centered medicine/health care and assistive technology.
- Disability awareness. People need to see positive examples of families and individuals coping with genetic disorders.
- The impact of a genetic condition in the family.
- How the Human Genome Project will provide scientific knowledge that may improve our daily lives in general.
- Ethical standards and genetic testing.

## Summary/Discussion

Priority topics for education on the ethical, legal, and psychosocial issues of the Human Genome Project for 332 consumers and 329 health professionals were presented. The responses came from two surveys conducted in 1994-95 by Georgetown University and the Alliance of Genetic Support Groups. The samples included 32 consumers and 25 health professionals who identified as African-American, Asian-American, American Indian, and Hispanic. Top priorities of consumers and providers, regardless of ethnicity/culture, were remarkably similar.

Although special outreach was made in newsletters and letters to genetic support groups to encourage participation of cultural minorities, the 32 who responded and participated in the survey under-represented the population at large. Possible reasons for under-representation of cultural minorities in the consumer population include: (1) Genetic support groups (which were the primary source for the consumer volunteers) are believed to be composed of members who are primarily female and Caucasian; (2) African-Americans may still be leery of genetic projects following the sickle cell screening project of the 1970's which resulted in persons who were carriers of sickle cell disease experiencing discrimination in insurance, employment, and other areas; (3) African-Americans may be concerned about programs funded by the U.S. government as a result of the decades long syphilis experiments in Macon County (Tuskegee) conducted by the U.S. Public Health Department; and (4) people in some cultures may consider talking about genetics a private matter to be discussed with family members and not with strangers on the telephone.

Possible reasons for under-representation in the health professional population is that they really are under-represented on the faculty of University Affiliated Programs across the country. The UAPs are working diligently in recruitment efforts to remedy this situation.

# CLASSROOM DEBATES: FACILITATING STUDENT AWARENESS OF HUMAN GENETICS

**Aleta Sullivan**

Although I am now teaching at the college level, I recently spent 13 years teaching grades 10 through 12 in a public high school—which was absolutely fascinating. I taught in a high school where the population was 80 percent African-American. I learned early on I had to approach things from a different standpoint and incorporate student points of view into my classroom. Instead of being just a lecturer—which I started out as, I became more of an experienced academic coach. What I mean is that, after going through an eight-week unit in human medical genetics, I started including different things into my classes to help the students develop decision-making skills so they could make their own decisions. I had to help them realize that my particular opinions on any subject had nothing to do with what they were to believe. They would ask me: "What do you think about abortion, Ms. Sullivan?" I would say: "No, no, no, I will not tell you. You have to make that decision on your own, and it is one that you have to make that you can live with."

One of the techniques that I used in my classroom was debates. My students debated topics such as Tay-Sachs disease, sickle cell anemia, and neurofibromatosis. We did this mainly because I have had students who have had all of those disorders or had them in their families. We had to deal with coping and, as a little classroom community, we did have a great deal of affection for each other. You do not hear that very much about public schools anymore, but there is a lot of love and affection in a good many public school situations.

Our Human Debate Unit was an interesting unit. It was not originally a biology unit; I believe it originally started in a 12th grade Shakespeare class. It struck me that I could use it. I would take a premise such as: two young people find out that they are sickle cell carriers who have decided to get married, and the pro and a con sides of that are debated. I chose sickle cell because there is a

The author is an instructor in Human Anatomy at the Pearl River Community College, Poplarville, MS 39470.

large population with sickle cell genes in the Hattiesburg, Mississippi area where I was teaching at the time. I have taught many sickle cell carriers who had no idea what it meant. I felt it was my job to help them both academically and psychologically deal with the ramifications of that. As I said, I have taught many children who have sickle cell anemia. In addition, the same group of people also had a predominance of polydactyly so that was a non-life-threatening trait to study. We talked about the old wives' tales because they knew how grandmama took care of that little extra finger on the side of their hand. She tied a string around it and let it rot off. That is how they took care of it.

Debates brought out the issues on both sides of the coin, no matter what the issue. It was an interesting situation. I had one young man, a 10th grader who was a very active church member. He was actually ordained as a preacher in his church, a very special young man. He was actively against abortion. I understood that completely having been in a religious situation all of my life with my father being a deacon. After we had gone through the debates and all of the different situations and also through a mock trial where we had the defense attorneys and defendants and a prosecuting attorney on the Baby Doe situation, he came up to me and said, "Ms. Sullivan, I now understand why abortions might be necessary." I said "No, don't make up your mind based on these two weeks of study. You need to live a little bit longer and study a little bit more and then make your decision. Even then you will have to remember it cannot just be an all-encompassing decision." I think he agreed with me. I never told him what I thought. But you can see that biology in a classroom is also very much philosophy in a classroom. And it is not my philosophy, but the philosophy the students develop themselves.

We talked about whether they knew a Down syndrome child. Most of them do. "They are so much fun; they are so sweet," my students said. We had the parents and Down syndrome children come to the class because we had them in our school situation. They realized that sometimes these are very special people, and an abortion during pregnancy might take from the parents a very special person. They looked at it from that point of view as well as the flip side of the coin. The amniocentesis procedure cannot tell you how mentally retarded the Down syndrome child will be or how physically affected the child will be— whether the stomach is connected to the lungs or whatever. All we know is that the child has an extra 21st chromosome. You can make a decision to abort the child and feel guilty the rest of your life, or you can maintain the pregnancy and then you might feel guilty about having put the child through that. Those were the things that were brought out in the classroom through the debates and through the mock trial. Extra reading was required for the students.

I used the "e" word in my class, too, that is, "evolution." Evolution is a favorite topic of mine. I also used "mutation" to mean change. That's all. I started with George Washington Carver and some of his plant experiments. Then we talked about mutations in plants. Some people think of the word "mutation" as being deleterious, but that is not so. In evolution, mutation is just the change in the DNA sequence. That is what it is. Whether for the good or for the bad. After studying a little botany first and changes in plants, I then moved into the study of human genetics and evolution. Change over time. That is what has taken place in our human population of diversity, our beautiful diversity.

My students were not necessarily considered academically talented, but this did not hamper us and I did not care. They learned a lot of biology, and they learned to think beyond themselves. As teenagers, the only thing that mattered to them was: me, me, me. They were very ethnocentric and egocentric, which goes back to the population genetics that Dr. Jackson referred to. I have read the same thing in Jerry Diamond's *Third Chimpanzee* which talks about the natural tendency of one group trying to elevate themselves above other groups. This is not necessarily right, but they do it—to make themselves feel better because they know their own deficiencies. Maybe that is part of it.

There are opportunities for teachers out there. I have been involved with a molecular biology project which was funded by the Cold Spring Harbor Laboratory (NY) and the National Science Foundation at the University of Southern Mississippi. I was "institutionalized" for six weeks in the summer from eight in the morning to nine at night. Our instructor was a workaholic. We learned all kinds of things like electrophoresis and how to incorporate that into a high school classroom. We learned how to pretend electrophoresis with food coloring and filter paper or paper towels, whatever you have, and do mock situations.

I was one of the first 100 people chosen for the Access Excellence program put out by Genentech which was really nice. I enjoyed that computer they gave me and the three years of America On-Line for free. I could communicate with other teachers who had the background and the experiences that I had had. I could search for activities on the Internet or even through America On-Line education forums and then use those in my classroom. I could get together in a chat room and share ideas with other teachers on how well a particular activity worked or didn't work. Two really good web sites for sharing and finding new materials and activities are the Access Excellence site at http://www.gene.com/ae/ and the Genetics Education Center at the University of Kansas Medical Center at http://www.kumc.edu/gec/

But with our present system, not enough high school teachers have the opportunity to attend conferences to learn what is going on and all of the different viewpoints. I think that is one thing we need most in our educational process. Because high school teachers are the people who can take it into the classrooms and break it down for students who won't be going on to college and hopefully equip the students to make decisions based on their knowledge. That is a hard thing to do and so many teachers have to do it. But I don't think teachers in other fields have to face that as much as a biology teacher does.

Speaking for teachers, I think we need to make sure more of them have opportunities to attend conferences and know about available opportunities so they can get out there and take advantage of them. There are things available for teachers and, yes, there is an apathy out there that makes some not apply for the programs. We need community and professional influence so that there is more out there for teachers so that they are able to get a feeling of confidence that they can succeed.

As I said, I have been involved in several projects. I went to Princeton last year with the Woodrow Wilson National Fellowship Program. I use myself as an example. A teacher might have a Southern accent, but she can still go to Princeton and learn some things and share some things that other people might find worthwhile.

# STUDENTS: HOW WE CAN HELP

## Carlton P. Jones

After I defended my thesis at Tuskegee University in the summer of 1995, I went off to a summer internship at the National Institutes of Health's National Center for Human Genome Research (now the National Human Genome Research Institute) in Bethesda, which is near my home in Silver Spring, Maryland. I always wanted to work at NIH. But while I was at North Carolina A&T, which is where I did my undergraduate work, I was not able to find the time. The summers just piled up on each other and the next thing I knew, I had graduated. When the opportunity came, I really jumped at the chance to work in human genome research. I considered it a tremendous opportunity to be able to get a summer internship there.

At Tuskegee, I was in animal science so you are probably wondering: "okay, how does someone in animal science, get into human science?" The thing is that science seems to encompass everything—especially in the field of biotechnology, which was my area of concentration at Tuskegee. The tools are the same, the techniques are the same. The only difference is the animal you are working with. One thing I noticed at NIH—there were not a lot of students of college age there. They were mostly high school students, which is good, but a lot of those high school students I was able to talk with were more interested in going into human medicine instead of into research. I have no problems with anyone going into veterinary medicine or human medicine, but I think young students should always look at all the options. Many of those at NIH were exceptional students, and I think they would probably be better suited for the big challenge of discovering the unknown that you find in research work. Research is all about making the unknown knowable.

As far as learning about the Human Genome Project, I do not think there are enough students who know about it. I think it is very important that they do because of the social issues that are involved—like life insurance. People need to be aware that big changes in science and technology are going to come upon

Carlton P. Jones received an M.S. in Animal Science from Tuskegee University in 1995 and is now a graduate student at Michigan State University.

us fairly rapidly. Perhaps some day we will be tracked or monitored by our DNA fingerprint instead of just our fingerprint. It is a scary thought that someday you may be denied insurance because you may carry a gene for some form of cancer. So it is very important that a lot of people become aware of the Human Genome Project and where it is going and its implications for the future. Even if not in science, everyone should be aware of current scientific knowledge, no matter how general. If politicians, for instance, are pushing for laws to make it legal for insurance companies to determine whether they accept or reject a person's application for insurance based on their DNA make-up, we all can make a very educated decision.

I would like to recommend strongly once again that students become involved in the Human Genome Project and, if there are students out there who are interested in doing science, be aware that every person addressed as "Dr." is not a medical doctor. He or she may have a doctorate in one of many fields. Some of the younger students do not know that. I think you need to start at an early age about being able to differentiate between the types of "Doctors." I like to tell students younger than myself to look at fields other than the traditional professions as they think about and plan for their futures. Even if students are not interested in pursuing a career in science, they should be aware of all the research going on. And they should get involved in fund-raising events for genetic research in such diseases as sickle cell and cancer.

As far as the benefits of technology are concerned, I would say that the benefits really outweigh any kind of fears that people have. I think it will eventually be possible to diagnose many different genetic diseases. It may one day become the reality to start preventative measures to correct the genetic defect before a newborn baby even leaves the hospital.

I would just like to conclude by saying that we as students need to play a more active role in understanding what the Human Genome Project is about, and then become more active in scientific discussions and projects that directly and indirectly benefit HGP research.

# The Role of the Church in Increasing Community Awareness of the HGP

## Edward L. Wheeler

In my presentation I will first seek to identify the significance of the church (or the synagogue or mosque) as a preserver and formulator of public values. Second, I will attempt to briefly describe some of the challenges that the Human Genome Project can expect to face in soliciting the support of the church in helping to make the community more aware of the work being done. Third, I want to suggest some ways the Human Genome Project might be successful in making its work available to the community through the church.

The church, with all its theological, cultural, ethnic and racial differences is a very powerful force of culture. Despite mixed information on the decline of membership in many mainline Protestant denominations and signs of a greater willingness on the part of some Catholics in the United States to question certain papal teachings, churches still command the respect, and in many cases, the devotion of people across the nation. The "Religious Right" has shown its power in the past 15 to 20 years to mobilize its constituency in shaping public policy on issues that affect all of us. The African-American Church covers a much more diverse theological and ideological spectrum, but it is still the most powerful institution in the African-American community. Every Sunday it commands the loyalty of millions of people across socio-economic, gender and age lines. The African-American Church provided the leadership and much of the following for the civil rights movement, and it can still inform, mobilize and empower people to address almost any issue affecting the African-American community.

The church is a powerful force in American society; however, it should be clearly understood that because of the church's close ties to American culture, it has a tendency to preserve and conserve the existing values of society rather than be a persistent force for change. To be sure, the church can be radical and prophetic at times, but its overall tendency, especially among the majority population, is to perpetuate societal values and morals. Martin Luther King, Jr.'s

The author, former Dean of the Chapel at Tuskegee University, was recently appointed president of Christian Theological Seminary, 1000 W. 42nd St., Indianapolis, IN 46208.

*Letter from a Birmingham Jail* speaks eloquently to this tendency.

The tendency toward preservation and conservation of societal values, however, need not be seen as inherently evil or wrong. The tendency simply points to the fact that the church as an institution embraces societal stability and gradual change as opposed to rapid change that most often challenges tradition and societal stability. Self-preservation is a reality for the church also.

The church's conservative nature forms a key ingredient in the challenge the Human Genome Project faces in soliciting the church's support in the effort to inform the public of the project's benefits. One component of that challenge is the historic reality that the church has been slow to embrace scientific insights that seem to cast suspicion on the church's interpretation of the biblical account of creation. Examples abound of the church's reluctance to accept such ideas as the sun being the center of our universe, and we continue to wrestle with the implications of Darwin's theory of evolution. For both Protestants and Catholics, the Bible is the essential record of our faith. Anything interpreted as a threat to its authority will be met with strong, persistent and at times irrational opposition.

A second component of the church's opposition is tied to the first. Whether intentional or not, scientists often seem unaware or unconcerned about the consequences of their work for religious communities and present their findings in such a way as to increase the opposition of the church community. There are scientists who present their theories with a certainty and arrogance that matches or exceeds the arrogance and assured self-righteousness of church leaders who also have a grasp of "The Truth." The resulting tensions only exasperate the suspicions that already exist between the two communities.

A third challenge to the Human Genome Project's attempt to solicit the church's support in informing the public comes specifically from the African-American community and other such communities that often find themselves minimalized and marginalized in America. Racism continues to be a reality in this nation, and it has permeated every aspect of our life. Unfortunately, the scientific community is not and has not been immune from its deadly presuppositions and erroneous conclusions. Everything from the "scientific" validation of the benefits of slavery and the syphilis experiments which took place right here in Macon County to the pseudo-science of *The Bell Curve* has created a suspicion that mitigates against the African-American church lending its support to scientific investigation. The prevailing attitude often seems to be, "if you do not respect my humanity, how can I trust your experiments?" The scientific community faces the very real challenge of overcoming years of racist practices and questionable ethics when it comes to African-Americans and non-Europeans in order to gain the trust of their religious institutions in the process of informing their communities of the benefits of the Human Genome Project.

Despite the very real challenges that exist and cannot be ignored, I believe the church can play an important role in informing its constituencies of the Human Genome Project's benefits for them and then soliciting their participation. The first process is to begin interpreting the HGP in a manner that confirms the best of the religious tradition's understanding of humanity and God's creative activity. Such an interpretation is perhaps best done in collaboration between scientists and theologians. For example, the wonder and beauty of human creation seems to be confirmed, not denied, by the discoveries of the Human Genome Project. Such a perspective could do much to lower the barriers between the church and the scientific community.

A second process in soliciting the church's support in informing people of the Human Genome Project is in making information available to key church leadership in laymen's terms and providing forums where they can have their questions answered. I presume this is the purpose of this conference and any publication that comes out of it. The targeted leadership should include pastors, denominational executives, and lay leaders. This process can be slow and rather tedious, but the benefits that can accrue from such meetings can produce results that are well beyond the work required.

If the church's leadership across denominational, ethnic, cultural and racial lines can be convinced of the significance and benefit of the Human Genome Project, they will become the ambassadors of the project in their own contexts. They can help parishioners appreciate the relationship between illness and genetic research and can become a resource for helping people cope with a diagnosis of a genetically-related disease. There is not enough money to pay for such advocates and supporters. At the same time, the religious community will also be among those speaking out when directions taken by the HGP or its social and ethical impacts appear harmful to their congregants and the common good of society.

A third consideration in securing the support of the church in the HGP is in increasing the number of African Americans and other underrepresented populations in every aspect of the work. Increasing the number of African-American scientists and institutions involved in the research will help allay some of the existing concerns about institutional racism. Involving such communities as Tuskegee University in HGP conferences should prove invaluable in increasing the credibility of the project in the eyes of the religious communities you seek to solicit for help. Increasing the involvement of African-American theologians and other underrepresented church leaders will help the HGP better appreciate the distinctions in perspectives that are essential to securing their support.

A fourth consideration in securing the support of the church in understanding and promoting the HGP lies in confession and repentance—two theological terms. Worlds of benefits await the Human Genome Project in securing the support of the church when it acknowledges its mistakes, recognizes legitimate concerns about its work by well-meaning people, and then clearly articulates its intention to keep the dialogue between the church and itself open. Not every church or religious community will accept the HGP, but I believe that a sizable majority can become partners with the scientific community in the HGP if the invitation is rightly extended.

The Human Genome Project is another marvelous example of the wonderful freedom human beings have to push back the curtain of ignorance and gain new insights on ourselves and our world. Such exploration does not require an alienation between the church and the scientific community. Both realms are essential to being fully human. To be sure, the two communities often approach the mysteries of life from different perspectives, but they need not be mutually exclusive or antagonistic. In fact they can provide a fuller picture of who we are as human beings when they can work together in a respectful relationship. I believe that the HGP provides a new opportunity to forge a cooperative relationship between the church and the scientific community. It is my prayer that we will not waste the opportunity.

# Appendices

# 1: PRIMER ON MOLECULAR GENETICS

## Introduction

The complete set of instructions for making an organism is called its genome. It contains the master blueprint for all cellular structures and activities for the lifetime of the cell or organism. Found in every nucleus of a person's many trillions of cells, the human genome consists of tightly coiled threads of deoxyribonucleic acid (DNA) and associated protein molecules, organized into structures called chromosomes (Figure 1).

If unwound and tied together, the strands of DNA would stretch more than 5 feet but would be only 50 trillionths of an inch wide. For each organism, the components of these slender threads encode all the information necessary for building and maintaining life, from simple bacteria to remarkably complex human beings. Understanding how DNA performs this function requires some knowledge of its structure and organization.

## DNA

In humans, as in other higher organisms, a DNA molecule consists of two strands that wrap around each other to resemble a twisted ladder whose sides, made of sugar and phosphate molecules, are connected by rungs of nitrogen-containing chemicals called bases. Each strand is a linear arrangement of repeating similar units called nucleotides, which are each composed of one sugar, one phosphate, and a nitrogenous base (Figure 2). Four different bases are present in DNA: adenine (A), thymine (T), cytosine (C), and guanine (G). The particular order of the bases arranged along the sugar-phosphate backbone is called the DNA sequence; the sequence specifies the exact genetic instructions required to create a particular organism with its own unique traits.

This article and its six figures are taken from the U.S. Department of Energy's *Primer on Molecular Genetics, 1992*. The *Primer* is available on the Internet at http://www.ornl.gov/hgmis/publicat/publications.html#primer. Also helpful as an introduction: DOE's 1996 publication, *To Know Ourselves*, 34 pp.; for copies, write Betty Mansfield, Managing Editor, *Human Genome News*, Oak Ridge National Laboratory, 1060 Commerce Park, MS 6480, Oak Ridge, TN 37830.

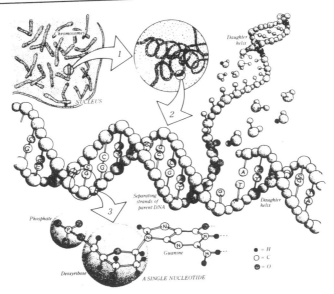

**Fig. 1. The Human Genome at Four Levels of Detail**

The two DNA strands are held together by weak bonds between the bases on each strand, forming base pairs (bp). Genome size is usually stated as the total number of base pairs; the human genome contains roughly 3 billion bp (Figure 3).

Each time a cell divides into two daughter cells, its full genome is duplicated; for humans and other complex organisms, this duplication occurs in the nucleus. During cell division the DNA molecule unwinds and the weak bonds between the base pairs break, allowing the strands to separate. Each strand directs the synthesis of a complementary new strand, with free nucleotides matching up with their complementary bases on each of the separated strands. Strict base-pairing rules are adhered to—adenine will pair only with thymine (an A-T pair) and cytosine with guanine

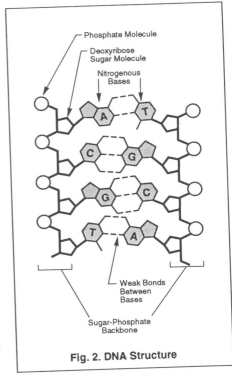

**Fig. 2. DNA Structure**

| COMPARATIVE SEQUENCE SIZES | (BASES) |
|---|---|
| (yeast chromosome 3) | 350 Thousand |
| *Escherichia coli* (bacterium) genome | 4.6 Million |
| Largest yeast chromosome now mapped | 5.8 Million |
| Entire yeast genome | 15 Million |
| Smallest human chromosome (Y) | 50 Million |
| Largest human chromosome (1) | 250 Million |
| Entire human genome | 3 Billion |

Fig. 3. Comparison of largest known DNA sequence with approximate chromosome and genome sizes of model organisms and Humans. A major focus of the Human Genome Project is the development of sequencing schemes that are faster and more economical.

(a C-G pair). Each daughter cell receives one old and one new DNA strand (Figures 1 and 4*).* The cell's adherence to these base-pairing rules ensures that the new strand is an exact copy of the old one. This minimizes the incidence of errors (mutations) that may greatly affect the resulting organism or its offspring.

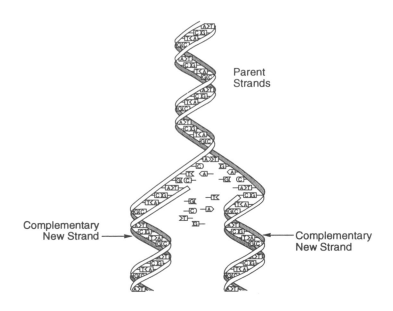

Parent
Strands

Complementary
New Strand

Complementary
New Strand

Fig. 4. DNA Replication

## Genes

Each DNA molecule contains many genes, the basic physical and functional units of heredity. A gene is a specific sequence of nucleotide bases, whose sequences carry the information required for constructing proteins which provide the structural components of cells and tissues as well as enzymes for essential biochemical reactions. The human genome is estimated to comprise at least 100,000 genes.

Human genes vary widely in length, often extending over thousands of bases, but only about 10% of the genome is known to include the protein-coding sequences (exons) of genes. Interspersed within many genes are intron sequences, which have no coding function. The balance of the genome is thought to consist of other noncoding regions (such as control sequences and intergenic regions), whose functions are obscure. All living organisms are composed largely of proteins; humans can synthesize at least 100,000 different kinds. Proteins are large, complex molecules made up of long chains of subunits called amino acids. Twenty different kinds of amino acids are usually found in proteins. Within the gene, each specific sequence of three DNA bases (codons) directs the cell's protein-synthesizing machinery to add specific amino acids. For example, the base sequence ATG codes for the amino acid methionine. Since 3 bases code for 1 amino acid, the protein coded by an average-sized gene (3000 bp) will contain 1000 amino acids. The genetic code is thus a series of codons that specify which amino acids are required to make up specific proteins.

The protein-coding instructions from the genes are transmitted indirectly through messenger ribonucleic acid (mRNA), a transient intermediary molecule similar to a single strand of DNA. For the information within a gene to be expressed, a complementary RNA strand is produced (a process called transcription) from the DNA template in the nucleus. This mRNA is moved from the nucleus to the cellular cytoplasm, where it serves as the template for protein synthesis. The cells protein-synthesizing machinery then translates the codons into a string of amino acids that will constitute the protein molecule for which it codes (Figure 5). In the laboratory, the mRNA molecule can be isolated and used as a template to synthesize a complementary DNA (cDNA) strand, which can then be used to locate the corresponding genes on a chromosome map. The utility of this strategy is described in the section on physical mapping.[1]

## Chromosomes

The 3 billion bp in the human genome are organized into 24 distinct, physically separate microscopic units called chromosomes. All genes are arranged linearly along the chromosomes. The nucleus of most human cells contains 2 sets of chromosomes, 1 set given by each parent. Each set has 23 single chromosomes, 22 autosomes and an X or Y sex chromosome. (A normal female will have a pair of X chromosomes; a male will have an X and Y pair). Chromosomes contain roughly equal parts of protein and DNA; chromosomal DNA

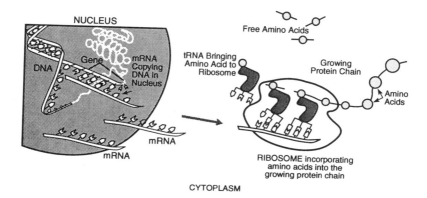

**Fig. 5. Gene Expression**

contains an average of 150 million bases. DNA molecules are among the largest molecules now known.

Chromosomes can be seen under a light microscope and, when stained with certain dyes, reveal a pattern of light and dark bands reflecting regional variations in the amounts of A and T vs G and C. Differences in size and banding pattern allow the 24 chromosomes to be distinguished from each other, an analysis called a karyotype. A few types of major chromosomal abnormalities, including missing or extra copies of a chromosome or gross breaks and rejoinings (translocations), can be detected by microscopic examination; Down syndrome, in which an individual's cells contain a third copy of chromosome 21, is diagnosed by karyotype analysis (Figure 6). Most changes in DNA, however, are

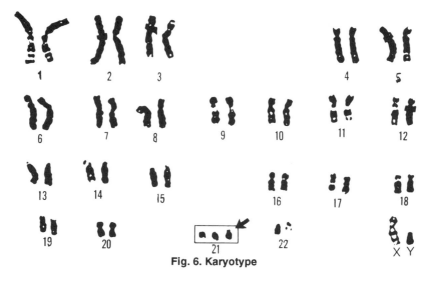

**Fig. 6. Karyotype**

too subtle to be detected by this technique and require molecular analysis. These subtle DNA abnormalities (mutations) are responsible for many inherited diseases such as cystic fibrosis and sickle cell anemia or may predispose an individual to cancer, major psychiatric illnesses, and other complex diseases.

**Editors' Note:** In addition to the above introductory information, this website can be accessed for information on: Mapping and Sequencing; Sequencing Technologies; End Games; Model Organisms; Informatics; and Impact. Additional websites (by no means complete) are listed and annotated beginning on page 277.

## 2: HEREDITARY COLON CANCER:
### Genetic Discoveries Offer Hope for Prevention

*Case 1:*  Beth M.'s father died of colon cancer, as did her grandmother. Now two of her brothers, both in their 40's, have been diagnosed with colon cancer. Beth, age 37, feels a curse is hanging over her family and is worried about her future and that of her children.

*Case 2:*  Paul C. was 35 when his doctor told him the grim news: he had advanced colon cancer. As far as he knew, Paul had no family history of the disease. But after checking, Paul learned that several aunts and uncles had died of colon cancer at an early age.

Further research revealed that some members of both Beth and Paul's families carry an altered gene, passed from parent to child, that predisposes them to a form of inherited colon cancer, known as hereditary nonpolyposis colorectal cancer (HNPCC). Sometimes difficult to diagnose, HNPCC is believed to account for one in six of all colon cancer cases. Cancers arise from a multi-step process, which involves the interplay of multiple changes, or mutations, in several different genes, in combination with environmental factors such as diet or lifestyle. In the most common, noninherited forms of cancer, the genetic changes are acquired after birth. But individuals who have an hereditary risk for cancer are born with one altered gene—in other words, they are born one step into the cancer process. In hereditary nonpolyposis colorectal cancer, for instance, children who inherit an altered gene from either parent face a 70 to 80% chance of developing this disease, usually at an early age. Women also face a markedly increased risk of uterine and ovarian cancer.

Though scientists had known for years that an altered gene was to blame for this hereditary colon cancer, finding it was tricky for they had few clues as

*Hereditary Colon Cancer* is taken from NIH Publication No. 95-3897 and can be accessed on the Internet at http://www.nhgri.nih.gov/Policy...and...public...affairs/Communications/Publications/ For more information about the Human Genome Project at the National Center for Human Genome Research at NIH, contact the Office of Communications, Rm.4B09, 9000 Rockville Pike, Bethesda, MD 20892 (301)-402-0911. For related websites, access http://www.nhgri.nih.gov.

to where, on any of the 23 pairs of chromosomes, the gene might reside. Finally, using tools emerging from the Human Genome Project, an international team tracked the gene to a region of chromosome 2. Seven months later, two teams zeroed in on the culprit. Just three months after that, they had identified a second gene on chromosome 3 also at work in HNPCC. Together, these genes account for most cases of this inherited cancer.

These discoveries offer a preview of how the Human Genome Project is likely to transform medicine by opening up new approaches to prevention. The earliest beneficiaries will be those families facing a very high risk of colon cancer. First, for those who choose to take it, will come a simple blood test to determine who in these cancer-prone families does or does not carry the altered genes. The consequences could be enormous, for as many as 1 in 200, or 1 million Americans, may carry one or the other of these altered genes. Individuals found to carry an altered gene would likely be counseled to adopt a high-fiber, low-fat diet in the hope of preventing cancer. They would also be advised to start yearly examinations of the colon at about age 30. Such exams should help physicians detect any benign polyps, wart-like growths on the colon, early in the disease process and then remove them before they turn malignant. For those individuals who turn out not to carry the altered genes, the diagnostic test may be a huge relief, removing the fear they have lived under and sparing them the need for frequent colonoscopies.

Despite the life-saving potential of such diagnostic tests, numerous issues need to be resolved before they are introduced into general medical practice. Genetic testing is not so simple as drawing blood and telling someone the results. For one, the best way to test large numbers of individuals is by no means clear. In deciding whether or not to be tested, individuals need information not only about the disorder and its risk but also about the test and its limitations. Equally important, genetic testing must be accompanied by counseling to help people cope with information about their future risk, whatever the outcome of the test. Those who test positive and who are trying to decide what course to pursue will need to know how effective various strategies, such as frequent colonoscopy and polyp removal, actually are at preventing colon cancer. Definitive answers are still lacking for these questions. Broader, societal issues arise as well, such as how to protect the confidentiality of genetic information and ensure that it is not used to discriminate against individuals in employment or insurance.

Even before these colon cancer susceptibility genes were discovered, the Human Genome Project had begun planning pilot studies to address these and other questions about testing for cancer risk. It is important that these questions be answered now, before widespread testing begins. The identification of genes involved in hereditary colon cancer is just one in a long string of discoveries that can be expected as the Human Genome Project progresses. Careful attention to these social and ethical issues now will help prepare the public and the medical profession for the choices that lie ahead.

# Appendix 3: USEFUL WEBSITES

## U.S. Government Agencies

*(1) http://www.ornl.gov/hgmis* (U.S. Department of Energy Human Genome Program) Quick fact finder; history, progress and goals of the HGP; support groups; DOE's role in the HGP; publications; glossary; answers to FAO's for students; ELSI information; other links.

*(2) http://www.nhgri.nih.gov* (National Human Genome Research Institute at the National Institutes of Health) Information about the HGP; intramural research; grants; policy and public affairs; genomic and genetic data and topical events.

*(3) http://ohrm.od.nih.gov* (National Institutes of Health) Personnel involved in biomedical sciences; developments in biomedicine, interactive learning about new technologies; other links.

*(4) http:www.lbl.gov/Education/Genome* (Genome Educators Homepage) Who's who in genome education, events calendar, event submission form, directory of educators.

## Genome Centers and Other Major Research Sites

*(1) http://www.kumc.edu/gec* (Genetics Education Center, University of Kansas Medical Center) Information about genetic resources (support groups, newsletters, genetic counseling, resource materials), the *HGP* (community and continuing education, weekly map updates, project description) and Networking (meeting calendars, educators' directory, mentor network, high school genome program network) may be accessed.

---

This partial list of websites that may be of interest to high school or college teachers was compiled and annotated by Sophia Ramlal, a third year student at Tuskegee University's School of Veterinary Medicine, with assistance from the U.S. Department of Energy and the National Institutes of Health.

*(2) http://www.ethics.ubc.ca/brynw/* (Center for Applied Ethics, University of British Columbia) *HGP/ELSI;* news and bulletins; conferences, public action and discussions groups, journals, genetics and the law; general philosophy.

*(3) http://raven.umnh.utah.edu/review/redo/newsmenu.html* (University of Utah) Topical information about research findings by University of Utah personnel.

*(4) http://www.med.upenn.edu/bioethics/* (University of Pennsylvania) Virtual library including publications authored by the faculty of the University of Pennsylvania; bioethics for beginners—introduction, organizations, help with homework, graduate program information; faculty members and contact information.

*(5) http://cbbridges.harvard.edu:7081* (Harvard University)
*http://flybase.bio.indiana.edu:82* (Indiana University) Both sites contain genetic and molecular data for *Drosophila*, representing part of the worldwide database for the study of this organism.

*(6) http://www-genome.stanford.edu/* (Stanford University) *Saccharomyces* and *Arabidopsis thaliana* databases; laboratory sequencing protocol manual; Macintosh software; internships for high-school students; nucleotide sequence search tool (BLAST); mapping; feedback.

*(7) http://www-leland.stanford.edu/~ger/drosophila.html* (Stanford University) *Drosophila* genome project; *Drosophila* labs on the net; the interactive fly; virtual FlyLab; archives; virtual library.

*(8) http://cc.emory.edu/PEDIATRICS/corn/corn.html* (Council of Regional Networks for Genetic Services at Emory University) Information about CORN, a Special Project of Regional and National Significance (SPRANS) sponsored by the USDHHS; Regional genetic network services; Alliance of Genetic Support Groups; goals, committees, officers and membership of "CORN."

*(9) http:eatworms.swmed.edu* (University of Texas Southwestern Medical Center) Recent papers; links to picture pages software; meetings related to *Caenorhabditis elegans,* a soil nematode used to study the genetics and development of neurobiology; other nematodes; virtual library.

*(10) http://hshgp.genome.washington.edu* (University of Washington) High school genome program web resource which facilitates trouble-shooting and problem-solving for teachers and students. It also provides two teaching modules—two in science and one in ethics, as well as a tutorial (virtual DNA sequencing) that lets students participate in parts of the sequencing process, including analyzing a sequencing ladder and carrying out a sequencing assembly.

## Other Institutions

*(1) http://agriculture.tusk.edu/caens/genome/genome.html* (Tuskegee University College of Agricultural, Environmental and Natural Sciences, Animal Genetics Laboratory) Information about ordering and educational discounts for *Plain Talk about the Human Genome Project,* the proceedings of the September, 1996 Tuskegee University Conference on the Human Genome Project. Additional information about the animal genetics laboratory operations, collaborators, publications, high school program, and current projects are also available.

*(2) http://darwin.cshl.org/facil.html* (Cold Spring Harbor's DNA Learning Center) Information about facilities, exhibits (Biomedia lab, multitorium, museum exhibit), workshops, laboratory field trip offerings, publications, products and student allele database.

*(3) http://www.gene.com/ae* (Access Excellence-Genentech) Scientific updates, on-line projects, technology in the classroom for the 21st century, science education reform, newsmakers, biotechnology partnerships, science seminars and a graphics gallery are some of the features offered through this site.

*(4) http://www.faseb.org/genetics/mainmenu.html* (American Society of Human Genetics) Listing of genetic societies; how to join; certification in genetics; genetics-related employment opportunities

*(5) http://gdb.org/gdbreports/CountGeneByChromosome.html* (Weekly updated count of mapped genes by chromosomes in tabular form.

*(6) http://www.infobiogen.fr/agora/eshg/* (The European Society of Human Genetics) Statutes; committees; list of annual meetings; courses; partnerships; on-line services for those with and without authentication; gopher service; *European Journal of Human Genetics.*

*(7) http: //www.faseb.org/genetics/webliog.html* Webliography for clinical geneticists; GENLINE-prototype electronic textbook for genetic conditions for which molecular testing is available; human gene map—genes featured are linked to articles of *Science Online;* HELIX; laboratories doing testing for genetic conditions are but a few of the sites listed.

*(8) http://www.applmeapro.com/agt/default.html* (The Association of Genetic Technologists) Membership; publications, inquiries, educational opportunities; annual meeting; other related organizations.

*(9) http://www.mcet.edu/humangenome* (Massachusetts Corporation for Educational Telecommunications) Designed for high school students and teachers,

this resource facilitates interactive student learning and showcasing of student talent through essay writing, drawing, painting, and photography. Dialogue about monthly topics is made possible through a Chat Room, while the Resource Center provides material useful as teaching aides for high school teachers.

**Note:** Four URLs that can be used to find human genome research sites anywhere in the world and learn what is going on are: (1) and (2) under U.S. Government Agencies above, the U. S. Department of Energy's HGP URL: http://www.er.doe.gov/production/ober/HELSRD_top.html and HUGO's (International Human Genome Organization) URL: http://www.hugo/gdb.org/

# GLOSSARY

**Adenine (A):** A nitrogenous base, one member of the *base pair\** A-T *(adenine-thymine)*.

**Alleles:** Alternative forms of a genetic *locus;* a single allele for each locus is inherited separately from each parent (e.g., at a locus for eye color the allele might result in blue or brown eyes).

**Amino acid:** Any of a class of 20 molecules that are combined to form *proteins* in living things. The sequence of amino acids in a protein and hence protein function are determined by the *genetic code.*

**Amplification:** An increase in the number of copies of a specific DNA fragment; can be *in vivo* or *in vitro*. See *cloning, polymerase chain reaction.*

**Arrayed library:** Individual primary recombinant clones (hosted in *phage, cosmid, YAC,* or other *vector)* that are placed in two-dimensional arrays in microtiter dishes. Each primary clone can be identified by the identity of the plate and the clone location (row and column) on that plate. Arrayed libraries of clones can be used for many applications, including screening for a specific gene or genomic region of interest as well as for *physical mapping.* Information gathered on individual clones from various genetic *linkage* and *physical map* analyses is entered into a relational database and used to construct physical and genetic *linkage maps* simultaneously; clone identifiers serve to interrelate the multilevel maps. Compare *library, genomic library.*

**Autoradiography:** A technique that uses X-ray film to visualize radioactively labeled molecules or fragments of molecules; used in analyzing length and number of DNA fragments after they are separated by gel *electrophoresis.*

**Autosome:** A *chromosome* not involved in sex determination. The *diploid* human *genome* consists of 46 chromosomes, 22 pairs of autosomes, and 1 pair of *sex chromosomes* (the X and Y chromosomes).

**Base pair (bp):** Two nitrogenous bases *(adenine* and *thymine* or *guanine* and *cytosine)* held together by weak bonds. Two strands of DNA are held together in the shape of a double helix by the bonds between base pairs.

**Base sequence:** The order of *nucleotide* bases in a DNA molecule.

**Base sequence analysis:** A method, sometimes automated, for determining the *base sequence.*

\*Italicized words within an explanation are listed in the Glossary.

The Glossary was taken from the U.S. Department of Energy's website that can be found at http://www.ornl.gov/hgmis/publicat/publications.html#glossary

**Biotechnology:** A set of biological techniques developed through basic research and now applied to research and product development. In particular, the use by industry of *recombinant DNA,* cell fusion, and new bioprocessing techniques.

**bp:** See *base pair.*

**cDNA:** See *complementary DNA.*

**Centimorgan (cM):** A unit of measure of *recombination* frequency. One centimorgan is equal to a 1% chance that a marker at one genetic *locus* will be separated from a marker at a second locus due to *crossing over* in a single generation. In human beings, 1 centimorgan is equivalent, on average, to 1 million *base pairs.*

**Chromosomes:** The self-replicating genetic structures of cells containing the cellular DNA that bears in its *nucleotide* sequence the linear array of *genes.* In *prokaryotes,* chromosomal DNA is circular, and the entire genome is carried on one chromosome. *Eukaryotic* genomes consist of a number of chromosomes whose DNA is associated with different kinds of *proteins.*

**Clone bank:** See *genomic library.*

**Clones:** A group of cells derived from a single ancestor.

**Cloning:** The process of asexually producing a group of cells (clones), all genetically identical, from a single ancestor. In *recombinant DNA technology,* the use of DNA manipulation procedures to produce multiple copies of a single *gene* or segment of DNA is referred to as cloning DNA.

**Cloning vector:** DNA molecule originating from a *virus,* a *plasmid,* or the cell of a higher organism into which another DNA fragment of appropriate size can be integrated without loss of the vector's capacity for self-replication; vectors introduce foreign DNA into host cells, where it can be reproduced in large quantities. Examples are *plasmids, cosmids,* and *yeast artificial chromosomes;* vectors are often *recombinant* molecules containing DNA sequences from several sources.

**Code:** See *genetic code.*

**Complementary DNA (cDNA):** DNA that is synthesized from a *messenger RNA* template; the single-stranded form is often used as a *probe* in *physical mapping.*

**Complementary sequences:** *Nucleic acid base sequences* that can form a double-stranded structure by matching *base pairs;* the complementary sequence to G-T-A-C is C-A-T-G.

**Conserved sequence:** A *base sequence* in a DNA molecule (or an *amino acid* sequence in a *protein)* that has remained essentially unchanged throughout evolution.

**Contig map:** A map depicting the relative order of a linked *library* of small overlapping clones representing a complete chromosomal segment.

**Contigs:** Groups of *clones* representing overlapping regions of a *genome.*

**Cosmid:** Artificially constructed *cloning vector* containing the *cos* gene of *phage* lambda. Cosmids can be packaged in lambda phage particles for infection into E. *coli;* this permits cloning of larger DNA fragments (up to 45 kb) than can be introduced into bacterial hosts in *plasmid* vectors.

**Crossing over:** The breaking during *meiosis* of one maternal and one paternal *chromosome,* the exchange of corresponding sections of DNA, and the rejoining of the chromosomes. This process can result in an exchange of alleles between chromosomes. Compare *recombination.*

**Cytosine (C):** A *nitrogenous base,* one member of the *base pair G-C (guanine* and *cytosine).*

**Deoxyribonucleotide:** See *nucleotide.*

**Diploid:** A full set of genetic material, consisting of paired *chromosomes* one chromosome from each parental set. Most animal cells except the *gametes* have a diploid set of chromosomes. The diploid human *genome* has 46 chromosomes. Compare *haploid.*

**DNA (deoxyribonucleic acid):** The molecule that encodes genetic information. DNA is a double-stranded molecule held together by weak bonds between *base pairs* of *nucleotides.* The four nucleotides in DNA contain the bases: *adenine* (A), *guanine (G), cytosine (C),* and *thymine (T).* In nature, *base pairs* form only between A and T and between *G* and C; thus the *base sequence* of each single strand can be deduced from that of its partner.

**DNA probes:** See *probe.*

**DNA replication:** The use of existing DNA as a template for the synthesis of new DNA strands. In humans and other *eukaryotes,* replication occurs in the cell *nucleus.*

**DNA sequence:** The relative order of *base pairs,* whether in a fragment of DNA, a *gene,* a *chromosome,* or an entire *genome.* See *base sequence* analysis.

**Domain:** A discrete portion of a *protein* with its own function. The combination of domains in a single protein determines its overall function.

**Double helix:** The shape that two linear strands of DNA assume when bonded together.

**E. coli:** Common bacterium that has been studied intensively by geneticists because of its small genome size, normal lack of pathogenicity, and ease of growth in the laboratory.

**Electrophoresis:** A method of separating large molecules (such as DNA fragments or *proteins)* from a mixture of similar molecules. An electric current is passed through a medium containing the mixture, and each kind of molecule travels through the medium at a different rate, depending on its electrical charge and size. Separation is based on these differences Agarose and acrylamide gels are the media commonly used for electrophoresis of proteins and nucleic acids.

**Endonuclease:** An *enzyme* that cleaves its nucleic acid substrate at internal sites in the *nucleotide* sequence.

**Enzyme:** A *protein* that acts as a catalyst, speeding the rate at which a biochemical reaction proceeds but not altering the direction or nature of the reaction.

**EST:** Expressed sequence tag. See *sequence tagged site.*

**Eukaryote:** Cell or organism with membrane-bound, structurally discrete *nucleus* and other well-developed subcellular compartments. Eukaryotes include all organisms except viruses, bacteria, and blue-green algae. Compare *prokaryote.* See chromosomes.

**Evolutionarily conserved:** See *conserved sequence.*

**Exogenous DNA:** DNA originating outside an organism.

**Exons:** The protein-coding DNA sequences of a *gene.* Compare *introns.*

**Exonuclease:** An *enzyme* that cleaves *nucleotides* sequentially from free ends of a linear nucleic acid substrate.

**Expressed gene:** See *gene expression.*

**FISH (fluorescence in situ hybridization):** A *physical mapping* approach that uses fluorescein tags to detect *hybridization* of *probes* with *metaphase chromosomes* and with the less-condensed *somatic interphase* chromatin.

**Flow cytometry:** Analysis of biological material by detection of the light-absorbing or fluorescing properties of cells or subcellular fractions (i.e., *chromosomes*) passing in a narrow stream through a laser beam. An absorbance or fluorescence profile of the sample is produced. Automated sorting devices, used to fractionate samples, sort successive droplets of the analyzed stream into different fractions depending on the fluorescence emitted by each droplet.

**Flow karyotyping:** Use of flow cytometry to analyze and/or separate *chromosomes* on the basis of their DNA content.

**Gamete:** Mature male or female reproductive cell (sperm or ovum) with a *haploid* set of *chromosomes (23* for humans). Gene: The fundamental physical and functional unit of heredity. A *gene* is an ordered sequence of *nucleotides* located in a particular position on a particular *chromosome* that encodes a specific functional product (i.e., a *protein* or *RNA molecule).* See *gene expression.*

**Gene expression:** The process by which a *genes* coded information is converted into the structures present and operating in the cell. Expressed genes include those that are transcribed into *mRNA* and then translated into *protein* and those that are transcribed into *RNA but* not translated into protein (e.g., *transfer* and *ribosomal RNAs).*

**Gene families:** Groups of closely related *genes* that make similar products.

**Gene library:** See *genomic library.*

**Gene mapping:** Determination of the relative positions of *genes* on a DNA molecule *(chromosome* or *plasmid)* and of the distance, in *linkage* units or physical units, between them.

**Gene product:** The biochemical material, either *RNA* or *protein,* resulting from expression of a gene. The amount of gene product is used to measure how active a gene is; abnormal amounts can be correlated with disease-causing alleles.

**Genetic code:** The sequence of *nucleotides,* coded in triplets *(codons)* along the *mRNA,* that determines the sequence of *amino acids* in *protein* synthesis. The DNA sequence of a *gene* can be used to predict the mRNA sequence, and the genetic code can in turn be used to predict the *amino acid* sequence.

**Genetic engineering technologies:** See *recombinant DNA technologies.*

**Genetic map:** See *linkage map.*

**Genetic material:** See *genome.*

**Genetics:** The study of the patterns of inheritance of specific traits.

**Genome:** All the genetic material in the *chromosomes* of a particular organism; its size is generally given as its total number of *base pairs.*

**Genome projects:** Research and technology development efforts aimed at *mapping* and *sequencing* some or all of the *genome* of human beings and other organisms.

**Genomic library:** A collection of *clones* made from a set of randomly generated overlapping DNA fragments representing the entire *genome* of an organism. Compare *library, arrayed library.*

**Guanine (G):** A nitrogenous base, one member of the *base pair* G-C (guanine and *cytosine*).

**Haploid:** A single set of *chromosomes* (half the full set of genetic material), present in the egg and sperm cells of animals and in the egg and pollen cells of plants. Human beings have 23 chromosomes in their reproductive cells. Compare *diploid*.

**Heterozygosity:** The presence of different *alleles* at one or more *loci* on *homologous chromosomes*.

**Homeobox:** A short stretch of *nucleotides* whose *base sequence* is virtually identical in all the *genes* that contain it. It has been found in many organisms from fruit flies to human beings. In the fruit fly, a homeobox appears to determine when particular groups of genes are expressed during development.

**Homologies:** Similarities in DNA or *protein* sequences between individuals of the same species or among different species.

**Homologous chromosomes:** A pair of *chromosomes* containing the same lmear *gene* sequences, each derived from one parent.

**Human gene therapy:** Insertion of normal DNA directly into cells to correct a genetic defect.

**Human Genome Initiative:** Collective name for several projects begun in 1986 by DOE to (1) create an ordered set of DNA segments from known chromosomal locations, (2) develop new computational methods for analyzing genetic map and DNA sequence data, and (3) develop new techniques and instruments for detecting and analyzing DNA. This DOE initiative is now known as the Human Genome Program. The national effort, led by DOE and NIH, is known as the Human Genome Project.

**Hybridization:** The process of joining two *complementary* strands of DNA or one each of DNA and RNA to form a double-stranded molecule.

**Informatics:** The study of the application of computer and statistical techniques to the management of information. In *genome* projects, informatics includes the development of methods to search databases quickly, to analyze DNA sequence information, and to predict *protein* sequence and structure from DNA sequence data.

**In situ hybridization:** Use of a DNA or RNA probe to detect the presence of the *complementary DNA* sequence in cloned bacterial or cultured *eukaryotic* cells.

**Interphase:** The period in the cell cycle when DNA is replicated in the nucleus; followed by *mitosis*.

**Introns:** The DNA *base sequences* interrupting the protein-coding sequences of a *gene;* these sequences are *transcribed* into *RNA* but are cut out of the message before it is *translated* into protein. Compare *exons*.

**In vitro:** Outside a living organism.

**Karyotype:** A photomicrograph of an individuals *chromosomes* arranged in a standard format showing the number, size, and shape of each chromosome type; used in low-resolution *physical mapping* to correlate gross chromosomal abnormalities with the characteristics of specific diseases.

**kb:** See *kilobase*.

**Kilobase (kb):** Unit of length for DNA fragments equal to 1000 *nucleotides*.

**Library:** An unordered collection of *clones* (i.e., cloned DNA from a particular organism), whose relationship to each other can be established by *physical mapping*. Compare *genomic library, arrayed library.*

**Linkage:** The proximity of two or more *markers* (e.g., *genes, RFLP* markers) on a *chromosome;* the closer together the markers are, the lower the probability that they will be separated during DNA repair or replication processes (binary fission in *prokaryotes, mitosis* or *meiosis* in *eukaryotes*), and hence the greater the probability that they will be inherited together.

**Linkage map:** A map of the relative positions of genetic *loci* on a *chromosome*, determined on the basis of how often the loci are inherited together. Distance is measured in *centimorgans (cM).*

**Localize:** Determination of the original position *(locus)* of a *gene* or other *marker* on a chromosome.

**Locus (pl.** loci): The position on a *chromosome* of a *gene* or other chromosome *marker;* also, the DNA at that position. The use of *locus is* sometimes restricted to mean regions of DNA that are *expressed.* See *gene*expression.

**Macrorestriction map:** Map depicting the order of and distance between sites at which *restriction enzymes* cleave *chromosomes.*

**Mapping:** See *gene mapping, linkage map, physical map.*

**Marker:** An identifiable physical location on a *chromosome* (e.g., *restriction enzyme cutting site, gene)* whose inheritance can be monitored. Markers can be expressed regions of DNA (genes) or some segment of DNA with no known coding function but whose pattern of inheritance can be determined. See *RFLP, restriction fragment length polymorphism.*

**Mb:** See *megabase.*

**Megabase (Mb):** Unit of length for DNA fragments equal to 1 million *nucleotides* and roughly equal to 1 cM.

**Meiosis:** The process of two consecutive cell divisions in the *diploid* progenitors of sex cells. Meiosis results in four rather than two daughter cells, each with a *haploid* set of *chromosomes.*

**Messenger RNA (mRNA):** RNA that serves as a template for protein synthesis. See genetic code.

**Metaphase:** A stage in *mitosis* or *meiosis* during which the *chromosomes* are aligned along the equatorial plane of the cell.

**Mitosis:** The process of nuclear division in cells that produces daughter cells that are genetically identical to each other and to the parent cell.

**mRNA:** See *messenger RNA.*

**Multifactorial or multigenic disorders:** See *polygenic disorders.*

**Multiplexing:** A *sequencing* approach that uses several pooled samples simultaneously, greatly increasing sequencing speed.

**Mutation:** Any heritable change in DNA *sequence.* Compare *polymorphism.*

**Nitrogenous base:** A nitrogen-containing molecule having the chemical properties of a base.

**Nucleic acid:** A large molecule composed of *nucleotide* subunits.

**Nucleotide:** A subunit of DNA or RNA consisting of a nitrogenous base *(adenine, guanine, thymine,* or *cytosine* in DNA; adenine, guanine, *uracil,* or cytosine in RNA), a phosphate molecule, and a sugar molecule (deoxyribose in DNA and ribose in RNA). Thousands of *nucleotides* are linked to form a DNA or RNA molecule. See *DNA, base pair, RNA.*

**Nucleus:** The cellular organelle in *eukaryotes* that contains the genetic material.

**Oncogene:** A *gene,* one or more forms of which is associated with cancer. Many oncogenes are involved, directly or indirectly, in controlling the rate of cell growth.

**Overlapping clones:** See *genomic library.*

**PCR:** See *polymerase chain reaction.*

**Phage:** A *virus* for which the natural host is a bacterial cell.

**Physical map:** A map of the locations of identifiable landmarks on DNA (e.g., *restriction enzyme cutting sites, genes),* regardless of inheritance. Distance is measured in *base pairs.* For the human *genome,* the lowest-resolution *physical map* is the banding patterns on the 24 different *chromosomes;* the highest-resolution map would be the complete *nucleotide* sequence of the chromosomes.

**Plasmid:** Autonomously replicating, extrachromosomal circular DNA molecules, distinct from the normal bacterial *genome* and nonessential for cell survival under nonselective conditions. Some plasmids are capable of integrating into the host genome. A number of artificially constructed plasmids are used as *cloning vectors.*

**Polygenic disorders:** Genetic disorders resulting from the combined action of *alleles* of more than one *gene* (e.g., heart disease, diabetes, and some cancers). Although such disorders are inherited, they depend on the simultaneous presence of several alleles; thus the hereditary patterns are usually more complex than those of *single-gene disorders.* Compare single-gene disorders.

**Polymerase chain reaction (PCR):** A method for amplifying a DNA *base sequence* using a heat-stable *polymerase* and two 20-base *primers,* one *complementary* to the (+)-strand at one end of the sequence to be amplified and the other complementary to the (-)-strand at the other end. Because the newly synthesized DNA strands can subsequently serve as additional templates for the same primer sequences, successive rounds of primer annealing, strand elongation, and dissociatlon produce rapid and highly specific amplification of the desired sequence. PCR also can be used to detect the existence of the defined sequence in a DNA sample.

**Polymerase, DNA or RNA:** *Enzymes* that catalyze the synthesis of *nucleic acids* on preexisting nucleic acid templates, assembling RNA from ribonucleotides or DNA from deoxyribonucleotides.

**Polymorphism:** Difference in DNA sequence among individuals. Genetic variations occurring in more than 1% of a population would be considered useful polymorphisms for genetic *linkage* analysis. Compare *mutation.*

**Primer:** Short preexisting polynucleotide chain to which new deoxyribonucleotides can be added by DNA *polymerase.*

**Probe:** Single-stranded DNA or RNA molecules of specific base *sequence,* labeled either radioactively or immunologically, that are used to detect the *complementary* base sequence by *hybridization.*

**Prokaryote:** Cell or organism lacking a membrane-bound, structurally discrete *nucleus* and other subcellular compartments. Bacteria are prokaryotes. Compare *eukaryote.* See *chromosomes.*

**Promoter:** A site on DNA to which *RNA polymerase* will bind and initiate *transcription.*

**Protein:** A large molecule composed of one or more chains of *amino acids* in a specific order; the order is determined by the *base sequence* of *nucleotides* in the *gene* coding for the protein. Proteins are required for the structure, function, and regulation of the bodys cells, tissues, and organs, and each protein has unique functions. Examples are hormones, *enzymes,* and antibodies.

**Purine:** A nitrogen-containing, single-ring, basic compound that occurs in nucleic acids. The purines in DNA and RNA are adenine and guanine.

**Pyrimidine:** A nitrogen-containing, double-ring, basic compound that occurs in nucleic acids. The pyrimidines in DNA are cytosine and thymine; in RNA, cytosine and uracil.

**Rare-cutter enzyme:** See *restriction enzyme cutting site.*

**Recombinant clones:** *Clones* containing *recombinant DNA molecules.* See *recombinant DNA technologies.*

**Recombinant DNA molecules:** A combination of DNA molecules of different origin that are joined using *recombinant DNA technologies.*

**Recombinant DNA technologies:** Procedures used to join together DNA segments in a cell-free system (an environment outside a cell or organism). Under appropriate conditions, a recombinant DNA molecule can enter a cell and replicate there, either autonomously or after it has become integrated into a cellular *chromosome.*

**Recombination:** The process by which progeny derive a combination of *genes* different from that of either parent. In higher organisms, this can occur by *crossing over.*

**Regulatory regions or sequences:** A DNA *base sequence* that controls *gene* expression.

**Resolution:** Degree of molecular detail on a *physical map* of DNA, ranging from low to high.

**Restriction enzyme, endonuclease:** A *protein* that recognizes specific, short *nucleotide sequences* and cuts DNA at those sites. Bacteria contain over 400 such *enzymes* that recognize and cut over 100 different DNA sequences. See *restriction enzyme cutting site.*

**Restriction enzyme cutting site:** A specific *nucleotide sequence* of DNA at which a particular *restriction enzyme* cuts the DNA. Some sites occur frequently in DNA (e.g., every several hundred *base pairs),* others much less frequently *(rare-cutter;* e.g., every 10,000 base pairs).

**Restriction fragment length polymorphism (RFLP):** Variation between individuals in DNA fragment sizes cut by specific *restriction enzymes; polymorphic sequences* that result in RFLPs are used as *markers* on both *physical maps* and genetic *linkage maps.* RFLPs are usually caused by *mutation* at a cutting site. See *marker.*

**RFLP:** See *restriction fragment length polymorphism.*

**Ribonucleic acid (RNA):** A chemical found in the *nucleus* and cytoplasm of cells; it plays an important role in protein synthesis and other chemical activities of the cell. The structure of RNA is similar to that of DNA. There are several classes of

RNA molecules, including *messenger RNA, transfer RNA, ribosomal RNA,* and other small RNAs, each serving a different purpose.

**Ribonucleotides:** See *nucleotide.*

**Ribosomal RNA (rRNA):** A class of RNA found in the ribosomes of cells.

**Ribosomes:** Small cellular components composed of specialized ribosomal RNA and protein; site of protein synthesis. See *ribonucleic acid (RNA).*

**RNA:** See *ribonucleic acid.*

**Sequence:** See *base sequence.*

**Sequence tagged site (STS):** Short (200 to 500 *base pairs)* DNA sequence that has a single occurrence in the human genome and whose location and base sequence are known. Detectable by *polymerase chain reaction, STSs* are useful for localizing and orienting the mapping and sequence data reported from many different laboratories and serve as landmarks on the developing *physical map* of the human genome. Expressed sequence tags (ESTs) are STSs derived from cDNAs.

**Sequencing:** Determination of the order of *nucleotides (base sequences)* in a DNA or *RNA* molecule or the order of *amino acids* in a *protein.*

**Sex chromosomes:** The X and Y *chromosomes* in human beings that determine the sex of an individual. Females have two X chromosomes in diploid cells; males have an X and a Y chromosome. The sex chromosomes comprise the 23rd chromosome pair in a *karyotype.* Compare *autosome.*

**Shotgun method:** *Cloning* of DNA fragments randomly generated from a *genome.* See *library, genomic library.*

**Single-gene disorder:** Hereditary disorder caused by a *mutant* allele of a single *gene* (e.g., Duchenne muscular dystrophy, retinoblastoma, sickle cell disease). Compare *polygenic disorders.*

**Somatic cells:** Any cell in the body except *gametes* and their precursors.

**Southern blotting:** Transfer by absorption of DNA fragments separated in electrophoretic gels to membrane filters for detection of specific *base sequences* by radiolabeled complementary probes.

**STS:** See *sequence tagged site.*

**Tandem repeat sequences:** Multiple copies of the same *base sequence* on a *chromosome;* used as a marker in *physical mapping.*

**Technology transfer:** The process of converting scientific findings from research laboratories into useful products by the commercial sector.

**Telomere:** The ends of *chromosomes.* These specialized structures are involved in the replication and stability of linear DNA molecules. See *DNA replication.*

**Thymine (T):** A nitrogenous base, member of *base pair* A-T (adenine-thymine).

**Transcription:** The synthesis of an *RNA* copy from a *sequence* of DNA (a *gene);* the first step in *gene expression.* Compare *translation.*

**Transfer RNA (tRNA):** A class of *RNA* having structures with triplet *nucleotide* sequences that are *complementary* to the triplet nucleotide coding sequences of *mRNA.* The role of tRNAs in protein synthesis is to bond with *amino acids* and transfer them to the ribosomes, where proteins are assembled according to the genetic code carried by mRNA.

**Transformation:** A process by which the genetic material carried by an individual cell is altered by incorporation of exogenous DNA into its *genome.*

**Translation:** The process in which the genetic code carried by mRNA directs the synthesis of *proteins* from amino acids. Compare *transcription.*

**tRNA:** See *transfer RNA.*

**Uracil:** A nitrogenous base normally found in RNA but not DNA; uracil is capable of forming a *base pair* with *adenine.*

**Vector:** See *cloning vector.* Virus: A noncellular biological entity that can reproduce only within a host cell. Viruses consist of *nucleic acid* covered by *protein;* some animal viruses are also surrounded by membrane. Inside the infected cell, the virus uses the synthetic capability of the host to produce progeny virus. VLSI: Very large-scale integration allowing over 100,000 transistors on a chip.

**YAC:** See *yeast artificial chromosome.*

**Yeast artificial chromosome (YAC):** A vector used to clone DNA fragments (up to 400 kb); it is constructed from the telomeric, centromeric, and replication origin sequences needed for replication in yeast cells. Compare *cloning vector, cosmid.*

# INDEX

abortion, 140, 152, 194, 221, 254
*ACE*, 41
Access Excellence Program, 255, 275
actuarial fairness, 161–163, 175–176, 185
adenine (A), in DNA, 1, 5
adverse selection *See* anti-selection
African-American, admixing, 99–100; Church, 259; Manifesto on Genomic Studies, 101
African-Americans, and the HGP, 215; common diseases of, 106–109; and sickle cell anemia, 143–146;
Agency for Health Care Policy and Research, 125
AIDs, *See* HIV
Alliance of Genetic Support Groups, 247
Alzheimer's disease, 113, 155, 193
American Academy of Actuaries, 169, 177, 184–185; Task Force on Genetic Testing, 169, 179
American College of Medical Genetics, 127
American Council of Life Insurance, 165
Americans with Disabilities Act, 115
American Society of Human Genetics, 91, 127, 275
American Society of Medical Genetics, 91
angiotensinogen, 39–40
anti-selection, in insurance, 171–172, 174, 189–184
Aristotle, 159, 161
Atomic Energy Commission, 2

autism, 199
autoimmune disease, 4, 13

Bacon, Francis, 117, 119, 121
BACs, 5, 19–20, 22, 63
Bannister, Roger, 221-222
basal cell carcinoma, 13
basal cell nevus syndrome, 13
base pairs (bp), 1, 266
base sequence, 17–18
behavior, and genetics, 85–91, 149, 167, 188, 216
*Bell Curve, The*, 54, 85, 89–91, 212, 260
Belmont Report, 79
beta thalasemmia, 69–70, 145, 218
biodiversity, 96
bioethics, 79–80, 117–118, 151, 156, 238
biological catalysts, 3
biological sciences, education model for, 225–241
bioprocessing, 3
bioremediation, 3
biotechnology, 60, 71–73, 120–121, 228, 233, 237–238, 240–241 245, 250; companies, 27, 114, 121, 135, 139–140, 188, 209, 225
Blue Cross/Blue Shield, 165
breast cancer, 6, 33, 113, 115, 128–129; case story, 117–120; familial implications, 119, 126; genes (*BRCA1*), 5, 27–29, 48, 141, 147, 155–156, 159, 164, 174, 176, 185, 188, 197, 213;